Amir Geranmayeh

Time Domain Boundary Integral Equations Analysis

Amir Geranmayeh

Time Domain Boundary Integral Equations Analysis

Stability, accuracy, and complexity concerns in 3D wave scattering; Application to wake field simulation in particle accelerators

Südwestdeutscher Verlag für Hochschulschriften

Imprint
Any brand names and product names mentioned in this book are subject to trademark, brand or patent protection and are trademarks or registered trademarks of their respective holders. The use of brand names, product names, common names, trade names, product descriptions etc. even without a particular marking in this work is in no way to be construed to mean that such names may be regarded as unrestricted in respect of trademark and brand protection legislation and could thus be used by anyone.

Publisher:
Südwestdeutscher Verlag für Hochschulschriften
is a trademark of
Dodo Books Indian Ocean Ltd., member of the OmniScriptum S.R.L Publishing group
str. A.Russo 15, of. 61, Chisinau-2068, Republic of Moldova Europe
Printed at: see last page
ISBN: 978-3-8381-2393-6

Zugl. / Approved by: Darmstadt, TU, Diss., 2010

Time Domain Boundary Integral Equations Analysis

Vom Fachbereich Elektrotechnik und Informationstechnik
der Technischen Universität Darmstadt

zur Erlangung des akademischen Grades
eines Doctor Ingenieurs (Dr.-Ing.)
genehmigte

DISSERTATION

von

M.Sc. Amir Geranmayeh
geboren am 14. September 1980 in Tehran

Darmstadt 2011

Referent:	Prof. Dr.-Ing. Thomas Weiland
Korreferent:	Prof. Dr.-Ing. Thomas Eibert

Tag der Einreichung:	20.10.2010
Tag der mündlichen Prüfung:	20.12.2010

D 17
Darmstadt 2011

To

Maryam Geranmaye

in Berlin

Contents

Acknowledgment

First and foremost, I wish to express my sincere gratitude and deepest appreciation to my PhD advisor, Prof. Thomas Weiland, whose abundant support has greatly helped me in the last five years. My probable future academic achievements surely would be beholden to his outstanding credibility. This work is heavily indebted to his broad vision in the field and the gracious credence he earnestly honored me.

I would like to cordially thank Prof. Thomas Eibert, Prof. Klaus Hofmann, Prof. Abdelhak M. Zoubir, and Prof. Gerd Balzer for the attention they devoted to review and evaluate my work as the examination committee.

Especial thanks should also go to Prof. Hideki Kawaguchi for providing suggestions courteously every year I met him.

I am deeply grateful to diligent mentor Dr. Wolfgang Ackermann whose ever-constant guidance learned me what the profound study through endless patience is all about and his scientific enthusiasm for performing pure flawless research tasks will remain as an utmost source of inspiration throughout my future career.

All my accomplishments would not have been possible without decades of steady encouragement and loving self-sacrifice of my adorable parents, Dr. Bahman and Tooba de Granmayeh.

And last but certainly not the least, I feel truly privileged to work with all affectionate friends and fellow colleagues who created such cheerful atmosphere in the TEMF institute; I indeed owe many TEMF graduate affiliates and the administrative assistants.

The financial support provided by the Deutschen Forschungsgemeinschaft (DFG) is highly acknowledged.

Abstract

The present research study mainly involves a survey of diverse time-domain boundary element methods that can be used to numerically solve the retarded potential integral equations. The aim is to address the late-time stability, accuracy, and computational complexity concerns in time-domain surface integral equation approaches. The study generally targets the transient electromagnetic scattering of three-dimensional perfect electrically conducting bodies. Efficient algorithms are developed to numerically solve the time-domain electric, derivative electric, magnetic, and combined field integral equation for the unknown induced surface current. The algorithms are mainly categorized into three major discretization schemes, namely the *marching-on-in-time*, the *marching-on-in-degree*, and the *convolution quadrature methods* or *finite difference delay modeling*. Possible choices of space-time integration are examined and the results are successfully compared with the high-resolution finite integration technique's solution to perform the converge study for practical applications where exact solutions are not available.

First consistent temporal interpolations with common time integrators are sought based on stability analysis of the *delay differential equations*. Besides, the higher orders of Lagrange and B-Spline time basis functions are employed to handle the time derivatives analytically. The orthogonal entire-domain but causal weighted Laguerre or Hermite polynomials are then employed to provide unconditionally stable *marching-on-in-degree* schemes. Moreover, the *convolution quadrature methods* which use a mapping from the Laplace domain to the z-domain based on the first or second finite difference approximation are investigated. In the *convolution quadrature methods*, the discretization is accomplished in the bilinear transform domain and the result is inverse transformed to create a time domain method in a marching style.

The outcome of this research study is applied to the non-dispersive modeling of the propagation of electromagnetic fields in particle accelerator structures, namely calculating the generated fields when the travelling bunches of charged particles passes through the beam line elements. The application of flexible and widely used Rao-Wilton-Glisson vector basis functions on flat triangular patches, particularly on the cylindrical beam pipes, causes in turn artificial fields in the commonly used barycentric approximation for the testing integrals due to misalignment of the surface normal vectors. To avoid such deficiency, first the cylindrical parts of the scatterers are supplanted by the rectangular ones whose unit normal vectors coincide with the real radial direction of the underlying cylindrical coordinate system. The linearly-varying divergence-conforming spatial basis functions on triangular and quadrilateral meshes are then combined. Additionally, in order to render symmetric interaction matrices complying the reciprocity theorem in the Galerkin's testing method while controlling the precision of numerical quadratures on the refining source and observation subdomains, the adaptive concurrent partitioning of planar patches is exploited. Furthermore, the eigenvalue spectrum of the system iteration matrix reveals that many stabilization techniques pull energy out of the system, and thus, symplectic space-time integration methods that fully conserve the energy are invoked.

A one-dimensional discrete fast Fourier transform-based algorithm is proposed to expedite the spatial convolution products of the Toeplitz-block-Toeplitz retarded interaction matrices. Additional saving owing to the system periodicity is linked with the Toeplitz properties due to the uniform space discretization in multi-level sense. In addition to the space-Fourier transformation algorithms, the time-Fourier transform routines are augmented to perform the recursive temporal convolution products for the Toeplitz block aggregates of the retarded interaction matrices in the outermost possible nested Toeplitz levels by array multiplications in spectral domain. Thus, the total computational cost and storage requirements scale down significantly in all the *marching-on-in-time* and *marching-on-in-degree* schemes or *convolution quadrature methods.* The temporal translation invariance properties of the time-tested Green's function are grouped in hybrid fixed and varying-size blocks to boost the efficiency of aggregate matrix-vector products in the diverse time-domain integral solver. Adaptive projection of triangular source elements on an auxiliary uniform grid is implemented for generalization of the algorithm to non-uniformly meshed scatterers. Novel summation reduction techniques are proposed to eliminate the most inner time-order loop in the *marching-on-in-degree* methods. Closed-form expression are presented for the discretized kernels when the *convolution quadrature methods* are applied for the time integration. Comparison of the exact near-field evaluation by the analytical integration on time-varying source subdomains with that of the polar integration is investigated as well. Cancelation of $\frac{1}{R^2}$ integrals in the *magnetic field integral equations* are explained to halve the computational cost of the *marching-on-in-time* schemes. It is shown that the solution procedure for several ten thousands spatial degrees of freedom and hundreds of time steps takes couple of days on a single quad-core machine.

Kurzfassung

Die vorliegende Arbeit beschäftigt sich mit der Erforschung verschiedener Formulierungen der Randelementemethode im Zeitbereich, die eingesetzt werden, um Integralgleichungen für retardierte Potentiale numerisch zu lösen. Ein wesentliches Ziel hier besteht darin, die Langzeitstabilität, Genauigkeit und Berechnungskomplexität für Integralgleichungsmethoden im Zeitbereich zu untersuchen. Die Studie zielt hauptsächlich darauf ab, das transiente Streuverhalten elektromagnetischer Felder für dreidimensionale, perfekt elektrisch leitfähige Körper zu beschreiben. Es werden effiziente Algorithmen zur numerischen Lösung der Integralgleichung auf Basis der elektrischen Feldstärke, der Zeitableitung der elektrischen Feldstärke, der magnetischen Feldstärke sowie einer Kombination dieser Felder hergeleitet, um die unbekannte induzierte Stromdichterverteilung zu berechnen. Die Algorithmen können in die drei Hauptkategorien *marching-on-in-time*, *marching-on-in-degree* und *convolution quadrature* Methoden bzw. *finite difference delay modeling* eingestuft werden. Mögliche Kombinationen von Raum- und Zeitintegrationen wurden untersucht und die erhaltenen Ergebnisse erfolgreich mit entsprechenden präzisen Lösungen auf Basis der Methode der Finiten Integration verglichen. Dies erlaubt die Durchführung von Konvergenzstudien auch für praktische Probleme, bei denen keine analytischen Lösungen vorliegen.

Auf der Basis einer Stabilitätsanalyse unter Verwendung der Theorie retardierter Differentialgleichungen wurden für die *marching-on-in-time* Verfahren zunächst konsistente Zeitinterpolationen zu den üblicherweise verwendeten Zeitintegratoren gesucht. Darüber hinaus wurden Lagrange- und B-Spline-Basisfunktionen höherer Ordnung eingesetzt, die eine Auswertung der benötigen Zeitableitungen auf analytischem Wege ermöglicht. Weiterhin wurden orthogonale, gewichtete Laguerre oder Hermite Polynome eingesetzt, die auf dem gesamten Gebiet kausal definiert sind, um stabile *marching-on-in-degree* Verfahren abzuleiten. Die *convolution quadrature* Methoden, die eine Abbildung von der Laplace- auf die z-Ebene ausnutzen, wurden unter Berücksichtigung sowohl von einseitigen als auch von zentralen Differenzenquotienten näher untersucht. Die Diskretisierung wird dabei in der Transformationsebene durchgeführt, wobei die Rücktransformation in den Zeitbereich ein stabiles Zeitschrittverfahren erzeugt.

Die Ergebnisse der Untersuchungen werden unter anderem zur dispersionsfreien Modellierung der Ausbreitung elektromagnetischer Felder in Teilchenbeschleunigern verwendet. Das Augenmerk liegt hier insbesondere in der Berechnung von resultierenden Feldern, die durch die Bewegung der geladenen Teilchenpakete durch die untersuchten Strahlführungselemente angeregt werden. Die sehr flexible und weit verbreitete Verwendung von geeigneten Basisfunktionen auf ebenen Dreiecksgittern führt insbesondere im Bereich der zylindrischen Strahlrohre zur unvermeidbaren Anregung von künstlichen Feldern, während sich diese auf entsprechenden rechteckigen Netzen vollständig vermeiden lassen. Im Übergangsbereich der Dreiecks- und der Rechtecksgitter müssen die Basisfunktionen dann divergenzkonform aufeinander angepasst werden. Um eine symmetrische Wechselwirkungsmatrix aufstellen zu können, welche die Reziprozität des zugrundeliegenden Galerkinverfahrens wiederspiegelt und eine vorgegebene Steuerung der Genauigkeit der numerischen Integration für das Quellen- und das Beobachtungsgebiet ermöglicht, wird gezielt eine adaptive simultane Zerlegung der jeweiligen Integrationsgebiete verwendet. Darüber hinaus wird gezeigt, dass viele der bekannten Stabilisierungsmaßnahmen einen Teil der im System vorhandenen Energie extrahiert und man deshalb nach einer Alternative sucht, die eine vollständige Erhaltung der Energie garantiert.

Zur Beschleunigung des räumlichen Faltungsprodukts der retardierten Wechselwirkungsmatrix in Toeplitz-Block-Toeplitz Gestalt wird ein schneller Algorithmus basierend auf der eindimensionalen diskreten schnellen Fouriertransformation vorgeschlagen. Durch eine einheitliche Diskretisierung des Raumes lassen sich zusätzliche Einsparungen erzielen, welche auf die Periodizität des Systems zurückzuführen ist und eng mit den Toeplitzeigenschaften des Systems in Verbindung steht. Zusätzlich zu den räumlichen Fourier-Transformations-Algorithmen wurden entsprechende zeitliche Routinen entwickelt, um die rekursiven zeitlichen Faltungsprodukte für die Toeplitzblöcke der retardierten Wechselwirkungsmatrix in den äußersten Toeplitzleveln durch Multiplikation der Matrizen im Spektralbereich auszuführen. Auf diese Weise kann sowohl der Rechen- als auch der Speicheraufwand für zeitliche und räumliche Freiheitsgrade in den *marching-on-in-time*, *marching-on-in-degree* und *convolution quadrature* Methoden reduziert werden. Die vorliegende Verschiebungsinvarianz der Greenfunktion in der Zeit erlaubt die Gruppierung in einzelne Blöcke mit sowohl fester als auch variabler Größe, um die Effizienz der Matrix-Vektor-Produkte in den verschiedenen Integrallösungsansätzen zu beschleunigen. Eine zusätzliche adaptive Projektion der Gitter ist implementiert, um die Algorithmen ebenfalls auf uneinheitlichen Gitter verwenden zu können. Durch eine Elimination der innersten Schleife wurde für die *marching-on-in-degree* Methoden weiterhin eine neue Möglichkeit gefunden, die notwendige Summation deutlich zu vereinfachen. Für die *convolution quadrature* Methoden wurden für die Zeitintegration geschlossene Lösungen präsentiert. Desweiteren wurde durch einen Vergleich der exakten Nahfeldauswertung mittels analytischer Integration über das Quellengebiet mit einer polaren Integration näher untersucht. Unter Verwendung der Auslöschung der $\frac{1}{R^2}$ Integrale in der *magnetic field integral equation* konnte eine Halbierung des Rechenaufwands für die *marching-on-in-time* Ansätze gezeigt werden. Weiterhin wurde demonstriert, dass die Berechnung der Lösung unter Verwendung von einigen zehntausend Unbekannten bei hunderten von Zeitschritten einige Tage Rechenzeit auf einer modernen Quad-Core-Workstation benötigt.

Chapter 1

Introduction

Applied computational electromagnetics seeks efficient and accurate algorithms for the solution of Maxwell equations' which indeed leads to the development of full-wave three-dimensional (3D) electromagnetic (EM[1]) simulation tools using numerical techniques for modeling wave radiation and propagation, inverse scattering problems, etc. It can ultimately provide better insight for engineers in broadband design, characterization, and optimization of (microstrip array) antennas, waveguides, dielectric resonators, high-frequency (microwave and millimeter waves) electrical circuits, radio communication systems and optical devices. It may also be used for interpreting radar signatures in remote sensing applications or estimating the shielding effectiveness against electromagnetic couplings and interferences.

1.1 Background and Motivations

In recent decades, the electromagnetics community has been pursued with renewed vigor the development of efficient transient simulators. Transient simulation of broadband electromagnetic radiations is commonly carried out based on solving either differential or integral equation type of the governing Maxwell equations. Since the surface integral equation-based techniques only require the discretization of the scatterer surface rather than the volume enclosing the whole structure in the finite difference time domain (FDTD) method, they intrinsically offer major advantages over the FDTD-based solvers when applied to the analysis of surface scatterers in homogeneous media. The integral equation approach also automatically imposes the radiation condition, and thus, there is no need for enforcing boundary conditions that are mostly required in the truncation of finite surrounding grids used by the FDTD method. Other advantageous of TDIE approaches over the versatile volume discretization methods is that no Courant-Friedrichs-Levy constrain on the time step size is needed in the former one. The TDIE are dispersion-free methods and as time-domain techniques, they analyze wide-band and potentially time-varying and nonlinear phenomena in a single simulation run.

Although time-domain boundary integral equation (TDIE) methods have long been known, they historically have been overlooked due to the high computational complexity and instabilities. In fact, when one chooses very small time steps to benefit from fast explicit methods, after a while unstable results appear before computing the system response for a sufficiently long period of time [1]-[11].

[1]The list of acronyms can be found in page 174.

The TDIE methods, conventionally solved by the marching-on-in-time (MOT) schemes, are increasingly receiving attention especially in the electromagnetic community for the complex broadband surface scattering phenomena and transient radiation problems [1]-[11]. The MOT schemes, computationally efficient time-domain formulation of the well-known boundary element method (BEM) for solving the TDIEs, have been shown prone to instabilities that appear in the form of exponentially growing oscillations in the late-time response and alternate in sign at each time step [12]-[24].The instabilities are originated at the system discretization stage in the conversion of the integral equation to a discrete time-space model. Historically, many authors have been postponed MOT instabilities through temporal filtering [1, 7, 13, 14, 16, 18, 19, 23]. The time averaging used in filterings, however, may adversely damage the accuracy of the final solution. The MOT instability arises when the poles that characterize the integral-equation system being solved drift into the right half-plane due to the approximations in the numerical scheme [13, 14]. Poles describing interior resonances permitted by the time-domain electric field integral equation (EFIE) and magnetic field integral equation (MFIE) are prime candidates for such undesirable shifts as they reside on the imaginary axis. A linear combination of the EFIE and MFIE, so-called the combined field integral equation (CFIE), has been exploited to eliminate the interior cavity modes that possibly corrupt the solution of the EFIE and MFIE [3, 4, 6]. Walker demonstrated that judiciously constructed MOT schemes designed to solve MFIE relying on accurate spatial integration rules and implicit time-stepping schemes are for practical purposes stable [2]. Nonetheless, precautions have to be taken in spatial [5] and temporal discretizations to avoid any pole displacement into the right half-plane. In respect to the spatial discretization, most researchers have used triangular patch modeling whereas all interior inner products of the BEM are evaluated using a fixed Gaussian quadrature rule over triangular subdomains, e.g. 7-points Gaussian-quadrature rule in [8]. Applying a predefined number of quadrature points, however, not only leads to a computationally inefficient algorithm for computing the interactions between those basis functions located far from each other, but also prohibit a sufficiently precise calculation of the mutual coupling of neighboring cells [25]. In other words, fix-point quadrature schemes prevent controlling over the precision of numerical integrations on a formerly meshed structure.

The computational cost associated with the application of the aforementioned marching schemes scales unfavorably with problem size. The inefficiency troubles of the MOT solvers recently have been addressed by the two-level [26] and multilevel plane-wave time-domain (PWTD) algorithms [27], accelerated Cartesian expansions (ACE) [28], and the hierarchical fast Fourier transform (FFT) methods [29], which derive their inspiration from the frequency-domain fast multipole and conjugate gradient FFT schemes, respectively. To eliminate the low-frequency breakdown and alleviate the ill-posedness of the surface EFIE operator when applied to densely discretized surfaces, Calderón multiplicative preconditioning [30, 32] and hierarchical regularization [33] have been proposed for the analysis of structures with sub-wavelength (electrically small) geometric features.

1.2 Advances Proposed by This Work

In respect to the temporal discretization, many authors have been yet favoured to employ either higher order difference formulas or smoother temporal interpolators in implicit time-stepping schemes so as to extend the span of the stable region [18, 23]. Such trends among electromagnetic engineers are being pursued by the traditional judgment based on the knowledge of the ordinary differential equations (ODE) whereas the retarded functional differential equations behave differently [34]. Mathematical foundations on the stability analysis of the delay differential equations (DDE) demonstrate that besides carefully designed spatial integration schemes, the conformity of the order of the time integration and the time interpolation is essential for the establishment of a stable TDIE-based solver [35].

In attempts to stabilize the MOT recipes, many researchers have also exploited diverse smooth time basis functions with less high-frequency content compared to the commonly used interpolators [19], only to interpolate the value of the retarded EM quantities between the past solution samples rather to employ their closed-form derivatives for time differentiation parts without corruption of the final results with any filtering. However, as mentioned earlier, the design of a new interpolator without considering the underlying integrator does not necessarily lead to a stable scheme. A wise approach to seek for well-matched integrator-interpolator combinations is checking the accuracy of schemes using approximating functions with analytical derivatives. The augmented exponential growth in the late-time regime of the system response, is principally originated at the system discretization stage, namely in the conversion of the integral equation to a discrete time-space model. Hence, employing continuous time basis functions to evaluate the time derivatives of the TDIE in closed form subsides numerical errors considerably and enhances the extent of stable region favorably [36]. Nonetheless, the marching-on-in-order, also referred as marching-on-in-degree (MOD), methods are yet the only approaches have thoroughly remedied the occurrence of the late-time instabilities whereby the time variations of the quantities are represented by the entire-domain damping Laguerre functions [10, 21].

This study aims to comprehensively review the recent advances in tackling the intrinsic late-time instabilities of the retarded potential integral equations appearing within the BEM. To this end, the thesis is organized as follows. Chapter 2 explains the derivation of kinds of time-domain surface integral equations for the analysis of EM radiation problems. Section 2.4 introduces a versatile numerical procedure decoupling the time evolution in a general variational form from the spatial integration in the time domain formulation of the BEM. In order to efficiently reach to the desired precision in spatial integrations, adaptive partitioning of triangular patches is proposed at the end of this chapter. Sections 3.2, 3.3, 3.4, and 3.5 describe four types of time discretization schemes on a similar platform for a stable solution of the EFIE, MFIE, and CFIE. First, consistent interpolants with various time integrators are explored based on the stability analysis of the DDE. As effective choices of subdomain time basis functions, different orders of the Lagrange and B-spline formulations with closed-form analytical derivatives are then investigated in Section 3.3. In Section 3.4, entire-domain orthogonal time basis functions, decaying to zero as the time progresses, are presented to guarantee the late-time stability of the BEM. Section 3.5 details temporal discretization mapping from the Laplace domain to the z-transform domain. The accuracy and convergence rate of the improved schemes are investigated based on comparison of the numerical results with the transient response obtained by employing the finite integration technique (FIT).

Chapter 2

Boundary Integral Equations

The electromagnetic (EM) fields are described by the time-dependent Maxwell's equations

$$\begin{aligned}
\nabla \times \mathbf{E}(\mathbf{r},t) &= -\frac{\partial}{\partial t}\mathbf{B}(\mathbf{r},t) \\
\nabla \times \mathbf{H}(\mathbf{r},t) &= \frac{\partial}{\partial t}\mathbf{D}(\mathbf{r},t) + \mathbf{J}_v(\mathbf{r},t) \\
\nabla \cdot \mathbf{D}(\mathbf{r},t) &= \rho(\mathbf{r},t) \\
\nabla \cdot \mathbf{B}(\mathbf{r},t) &= 0
\end{aligned} \tag{2.1}$$

in which the current $\mathbf{J}_v(\mathbf{r},t)$ as well as the electric and magnetic flux densities $\{\mathbf{D}(\mathbf{r},t), \mathbf{B}(\mathbf{r},t)\}$ are related to the field intensities $\{\mathbf{E}(\mathbf{r},t), \mathbf{H}(\mathbf{r},t)\}$ through linear[1], isotropic, homogeneous material properties (ϵ, μ, σ)

$$\begin{aligned}
\mathbf{D}(\mathbf{r},t) &= \epsilon\mathbf{E}(\mathbf{r},t) \\
\mathbf{B}(\mathbf{r},t) &= \mu\mathbf{H}(\mathbf{r},t) \\
\mathbf{J}_v(\mathbf{r},t) &= \sigma\mathbf{E}(\mathbf{r},t) + \mathbf{J}_0(\mathbf{r},t)
\end{aligned}$$

The EM fields in the coupled partial differential equations (PDE) (2.1) can be evaluated at any arbitrary space point $\mathbf{r}$ and time instance t by discretizing the volume encompassing the objects and calculating the fields on the surrounding grids via time-stepping [39]. Alternatively, the transient simulation of (broadband) EM radiations can be carried out based on solving the integral equation type of the governing Maxwell equations in which only the scatterer surface is needed to be discretized to find the induced surface current density $\mathbf{J}(\mathbf{r},t)$.

Let S denote the surface of an open or closed perfectly conducting body that resides in free space and that is illuminated by a transient EM field $\{\mathbf{E}^i(\mathbf{r},t), \mathbf{H}^i(\mathbf{r},t)\}$ as posed in Fig. 2.1. The interaction of the incident field with S results in a surface current density $\mathbf{J}(\mathbf{r},t)$, which in turn generates scattered EM fields $\{\mathbf{E}^s(\mathbf{r},t), \mathbf{H}^s(\mathbf{r},t)\}$. These fields can be fully characterized by the magnetic vector potential $\mathbf{A}(\mathbf{r},t)$ and the electric scalar potential $\phi(\mathbf{r},t)$

$$\mathbf{E}^s(\mathbf{r},t) = -\frac{\partial \mathbf{A}(\mathbf{r},t)}{\partial t} - \nabla\phi(\mathbf{r},t) \tag{2.2}$$

$$\mathbf{H}^s(\mathbf{r},t) = \frac{1}{\mu}\nabla \times \mathbf{A}(\mathbf{r},t) \tag{2.3}$$

[1]Hints for the extension to nonlinear or anisotropic inhomogeneous media can be found respectively in [37], [38].

where the potentials are defined as

$$\begin{aligned}\mathbf{A}(\mathbf{r},t) &= \frac{\mu}{4\pi}\int_S \mathbf{J}(\mathbf{r}',t) * G(R,t)\mathrm{d}S' \\ \phi(\mathbf{r},t) &= \frac{1}{4\pi\epsilon}\int_S \sigma(\mathbf{r}',t) * G(R,t)\mathrm{d}S'.\end{aligned}$$

Here, $R = |\mathbf{R}| = |\mathbf{r}-\mathbf{r}'|$ is the distance between the observation point $\mathbf{r}$ and the source point $\mathbf{r}'$ (arbitrarily located points on the surface S), the parameters μ and ϵ are the permeability and permittivity of the surrounding environment, the function G is the integral kernel, and $*$ represents the temporal convolution. Considering the free-space Green's function in 3D $G(R,t) = \frac{\delta(t-R/c)}{R}$, the potentials can be written as

$$\begin{aligned}\mathbf{A}(\mathbf{r},t) &= \frac{\mu}{4\pi}\int_S \frac{\mathbf{J}(\mathbf{r}',\tau)}{R}\mathrm{d}S' \\ \phi(\mathbf{r},t) &= \frac{1}{4\pi\epsilon}\int_S \frac{\sigma(\mathbf{r}',\tau)}{R}\mathrm{d}S'.\end{aligned} \tag{2.4}$$

The variable $\tau = t - R/c$ is called the retarded time and $\sigma(\mathbf{r},t)$ denotes the surface charge density which is related to the surface current density $\mathbf{J}(\mathbf{r},t)$ according to the continuity equation,

$$\nabla \cdot \mathbf{J}(\mathbf{r},t) + \partial_t\sigma(\mathbf{r},t) = 0, \tag{2.5}$$

and thus,

$$\phi(\mathbf{r},t) = -\frac{1}{4\pi\epsilon}\int_S\int_{-\infty}^{\tau} \frac{\nabla_{\mathbf{r}'} \cdot \mathbf{J}(\mathbf{r}',t')}{R}\mathrm{d}t'\mathrm{d}S'. \tag{2.6}$$

The retarded potentials ($\mathbf{A}$, ϕ) satisfy the Lorenz gauge condition on (2.1)

$$\nabla \cdot \mathbf{A} + \mu\epsilon\frac{\partial}{\partial t}\phi + \mu\sigma\phi = 0$$

and also the differential equations

$$\begin{aligned}\nabla^2\mathbf{A} - \mu\epsilon\frac{\partial^2}{\partial t^2}\mathbf{A} - \mu\sigma\frac{\partial}{\partial t}\mathbf{A} = \frac{\rho}{\epsilon} \\ \nabla^2\phi - \mu\epsilon\frac{\partial^2}{\partial t^2}\phi - \mu\sigma\frac{\partial}{\partial t}\phi = -\mu\mathbf{J}_0\end{aligned}$$

When the body S resides in a lossy homogeneous background with constant conductivity σ, according to the scalar telegraphers equation [40] a diffusive wake term is added to the scalar Green's function, i.e.,

$$G(R,t) = \frac{\exp(-\frac{\sigma}{2\epsilon}t)}{4\pi}\left[\frac{\delta(t-\frac{R}{c})}{R} + \frac{\sigma\eta \mathrm{I}_1\left(\frac{\sigma}{2\epsilon}\sqrt{t^2-(\frac{R}{c})^2}\right)}{2\sqrt{t^2-(\frac{R}{c})^2}}\mathrm{u}(t-\frac{R}{c})\right]$$

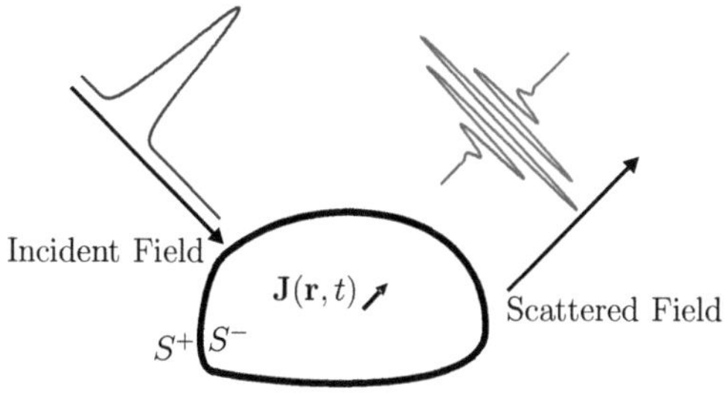

Figure 2.1: Description of a typical EM scattering problem; A known field is traveling towards the body wit surface S; S^+ and S^- denote slightly exterior and interior surfaces conformal to S residing just outside and just inside S, respectively.

where $\mathrm{I}_1(.)$ is the first-order modified Bessel function of the first kind. In 2D EM scattering, only the Green's function is replaced by

$$G(R,t) = \frac{\eta}{2\pi c}\frac{\mathrm{u}(t-\frac{R}{c})}{\sqrt{t^2-\left(\frac{R}{c}\right)^2}}$$

where $c = \frac{1}{\sqrt{\mu\epsilon}}$ is the velocity of the propagation of the EM wave in the space, $\mathrm{u}(.)$ is the Heaviside step function, and the intrinsic impedance of the space $\eta = \sqrt{\frac{\mu}{\epsilon}}$.

In order to set up the time-domain EFIE formulation, the incident field $\mathbf{E}^i(\mathbf{r},t)$ is related to the scattered field $\mathbf{E}^s(\mathbf{r},t)$ by enforcing the boundary condition: the total electric field $\mathbf{E}^t(\mathbf{r},t) = \mathbf{E}^i(\mathbf{r},t) + \mathbf{E}^s(\mathbf{r},t)$ tangential to S has to vanish, i.e.,

$$\begin{aligned} &\hat{\mathbf{n}} \times \left(\hat{\mathbf{n}} \times \mathbf{E}^t(\mathbf{r},t)\right) = 0, \\ &\hat{\mathbf{n}} \times \left(\hat{\mathbf{n}} \times \mathbf{E}^i(\mathbf{r},t)\right) = -\hat{\mathbf{n}} \times \left(\hat{\mathbf{n}} \times \mathbf{E}^s(\mathbf{r},t)\right) \qquad \forall \mathbf{r} \in S, \end{aligned} \tag{2.7}$$

where $\hat{\mathbf{n}}$ denotes an outward-directed unit vector normal to S at observation point $\mathbf{r}$.

Likewise, the time-domain MFIE can be derived by enforcing the condition that either the total magnetic field $\mathbf{H}^t(\mathbf{r},t) = \mathbf{H}^i(\mathbf{r},t) + \mathbf{H}^s(\mathbf{r},t)$ tangential to S equals the induced surface current density,

$$\hat{\mathbf{n}} \times [\mathbf{H}^i(\mathbf{r},t) + \mathbf{H}^s(\mathbf{r},t)] = \mathbf{J}(\mathbf{r},t) \qquad \forall \mathbf{r} \in S^+ \tag{2.8}$$

or the total magnetic field tangential to the inside surface of S vanishes

$$\hat{\mathbf{n}} \times \mathbf{H}^i(\mathbf{r},t) = -\hat{\mathbf{n}} \times \mathbf{H}^s(\mathbf{r},t) \qquad \forall \mathbf{r} \in S^-.$$

2.1 Electric Field Integral Equation (EFIE)

Substitution of (2.2) in (2.7) reads

$$\left.\frac{\partial \mathbf{A}(\mathbf{r},t)}{\partial t} + \nabla\phi(\mathbf{r},t)\right|_{\tan} = \hat{\mathbf{n}} \times \left(\hat{\mathbf{n}} \times \mathbf{E}^i(\mathbf{r},t)\right) \qquad \forall \mathbf{r} \in S \tag{2.9}$$

where the subscript "tan" denotes the tangential component. The excitation term on the right side of the EFIE (2.9), is shown by $\underline{\mathbf{E}}^i(\mathbf{r}, t)$ for the rest of the script, implying the incident electric field tangential to the surface S. Inserting (2.4) and (2.6) into (2.9) gives

$$\frac{\mu}{4\pi}\frac{\partial}{\partial t}\int_S \frac{\mathbf{J}(\mathbf{r}', \tau)}{R}\mathrm{d}S'\bigg|_{\tan} - \frac{\nabla_\mathbf{r}}{4\pi\epsilon}\int_S\int_{-\infty}^{\tau} \frac{\nabla_{\mathbf{r}'}\cdot\mathbf{J}(\mathbf{r}', t')}{R}\mathrm{d}t'\mathrm{d}S'\bigg|_{\tan} = \underline{\mathbf{E}}^i(\mathbf{r}, t) \tag{2.10}$$

which has to be solved for the unknown induced surface current density $\mathbf{J}(\mathbf{r}, t)$.

2.1.1 Alternative Forms of TD-EFIE

The derivative counterpart of the EFIE (DEFIE) is also of interest to preferably avoid the laborious computation of charge accumulation in solving the original form of EFIE (2.10). The DEFIE is obtained by taking a time derivative from both sides of (2.2) and (2.7)

$$\frac{\partial \mathbf{E}^s(\mathbf{r}, t)}{\partial t} = -\frac{\partial^2 \mathbf{A}(\mathbf{r}, t)}{\partial^2 t} + c^2\nabla\nabla\cdot\mathbf{A}(\mathbf{r}, t)$$

and combining them together

$$\frac{\partial^2 \mathbf{A}(\mathbf{r}, t)}{\partial t^2} - c^2\nabla\nabla\cdot\mathbf{A}(\mathbf{r}, t) = \frac{\partial \underline{\mathbf{E}}^i(\mathbf{r}, t)}{\partial t},$$

$$\frac{\mu}{4\pi}\int_S \left(\frac{\partial^2}{\partial t^2}\mathrm{I} - c^2\nabla\nabla\right)\cdot\frac{\mathbf{J}(\mathbf{r}', \tau)}{R}\mathrm{d}S' = \frac{\partial \underline{\mathbf{E}}^i(\mathbf{r}, t)}{\partial t}$$

where I is identity dyadic tensor in three dimensions. Alternatively, one may take the time derivative from both sides of (2.10),

$$\frac{\mu}{4\pi}\frac{\partial^2}{\partial t^2}\int_S \frac{\mathbf{J}(\mathbf{r}', \tau)}{R}\mathrm{d}S' - \frac{\nabla_\mathbf{r}}{4\pi\epsilon}\int_S \frac{\nabla_{\mathbf{r}'}.\mathbf{J}(\mathbf{r}', \tau)}{R}\mathrm{d}S' = \frac{\partial \underline{\mathbf{E}}^i(\mathbf{r}, t)}{\partial t} \tag{2.11}$$

in which the second term is called the Hertz potential

$$\Phi(\mathbf{r}, t) = \frac{\partial}{\partial t}\phi(\mathbf{r}, t) = -\frac{\nabla_\mathbf{r}}{4\pi\epsilon}\int_S \frac{\nabla_{\mathbf{r}'}.\mathbf{J}(\mathbf{r}', \tau)}{R}\mathrm{d}S'. \tag{2.12}$$

Taking the $\nabla_\mathbf{r}$ into the integral, according to (2.5), one needs the vector identity

$$\nabla_\mathbf{r}\left(\frac{\dot{\sigma}}{R}\right) = \frac{1}{R}\nabla_\mathbf{r}\dot{\sigma} + \nabla_\mathbf{r}\left(\frac{1}{R}\right)\dot{\sigma}, \qquad \nabla_\mathbf{r}\left(\frac{1}{R}\right) = -\frac{1}{R^2}\frac{\mathbf{R}}{R} \tag{2.13}$$

where the dot represent ∂_t and the gradient of the time-derivative of the retarded charge density is obtained using,

$$\nabla_\mathbf{r}\dot{\sigma}(\mathbf{r}', \tau) = \frac{1}{c}\frac{\partial}{\partial t}\dot{\sigma}(\mathbf{r}', \tau)\frac{\mathbf{R}}{R}. \tag{2.14}$$

Therefore, the gradient of the integrand in (2.13) can be represented by

$$\nabla_\mathbf{r}\frac{\dot{\sigma}(\mathbf{r}', \tau)}{R} = \frac{1}{c}\frac{\partial}{\partial t}\dot{\sigma}(\mathbf{r}', \tau)\frac{\mathbf{R}}{R^2} + \dot{\sigma}(\mathbf{r}', \tau)\frac{\mathbf{R}}{R^3}. \tag{2.15}$$

Finally, the combination of (2.11) and (2.15) yields in

$$\frac{1}{c^2}\frac{\partial^2}{\partial t^2}\int_S \frac{\mathbf{J}(\mathbf{r}',\tau)}{R}\mathrm{d}S' + \int_S \left[\frac{R}{c}\frac{\partial}{\partial t}+1\right]\nabla_{\mathbf{r}'}.\mathbf{J}(\mathbf{r}',\tau)\frac{\mathbf{R}}{R^3}\mathrm{d}S' = \frac{4\pi}{c\eta}\frac{\partial \underline{\mathbf{E}}^i(\mathbf{r},t)}{\partial t} \tag{2.16}$$

in which due to the time variation, the second integral can not be written as

$$\frac{1}{c}\frac{\partial^2}{\partial t^2}\int_S \frac{\mathbf{J}(\mathbf{r}',\tau)}{R}\mathrm{d}S' + c\int_S \nabla_{\mathbf{r}'}.\mathbf{J}(\mathbf{r}',\tau)\nabla_{\mathbf{r}}\left(\frac{1}{R}\right)\mathrm{d}S' = \frac{4\pi}{\eta}\frac{\partial \underline{\mathbf{E}}^i(\mathbf{r},t)}{\partial t}.$$

The DEFIE (2.11) calls for derivation of the transient excitation that may not be numerically realizable for very sharp impulsive pulses [41, 42]. Hence, the electric current can also be expressed in terms of the Hertz vector $\mathbf{c}(\mathbf{r},t)$, defined by $\mathbf{J}(\mathbf{r},t) = \frac{\partial}{\partial t}\mathbf{c}(\mathbf{r},t)$, whereby (2.10) can be written as

$$\frac{\mu}{4\pi}\frac{\partial^2}{\partial t^2}\int_S \frac{\mathbf{c}(\mathbf{r}',\tau)}{R}\mathrm{d}S' - \frac{\nabla_{\mathbf{r}}}{4\pi\epsilon}\int_S \frac{\nabla_{\mathbf{r}'}.\mathbf{c}(\mathbf{r}',\tau)}{R}\mathrm{d}S' = \underline{\mathbf{E}}^i(\mathbf{r},t). \tag{2.17}$$

As a result, the problem can be solved for the unknown Hertz vector instead of direct solving for the unknown surface current [10, 43]. Evidently, the Hertz approaches demand extra post-processing stages for computation of some desired electromagnetic quantities. Excluding the time derivation on the excitation term in (2.11), any computational model furnishing the numerical solution of (2.11) can be directly applied to (2.17) with identical excitation to the right-hand side of (2.10) [21].

2.2 Magnetic Field Integral Equation (MFIE)

Inserting (2.4) in (2.3) and combining the result with (2.8) gives

$$\mathbf{J}(\mathbf{r},t) = \hat{\mathbf{n}}\times\mathbf{H}^i(\mathbf{r},t) + \hat{\mathbf{n}}\times\left(\nabla\times\frac{1}{4\pi}\int_S \frac{\mathbf{J}(\mathbf{r}',\tau)}{R}\mathrm{d}S'\right). \tag{2.18}$$

2.2.1 Simplifications of the MFIE

Extracting the Cauchy principle value (PV) from the curl term, one can write the second term on the right-hand side of (2.18) as

$$\hat{\mathbf{n}}\times\left(\nabla\times\frac{1}{4\pi}\int_S \frac{\mathbf{J}(\mathbf{r}',\tau)}{R}\mathrm{d}S'\right) = \frac{\mathbf{J}(\mathbf{r},t)}{2} + \hat{\mathbf{n}}\times\left(\nabla\times\frac{1}{4\pi}\int_{S_0} \frac{\mathbf{J}(\mathbf{r}',\tau)}{R}\mathrm{d}S'\right) \tag{2.19}$$

where S_0 denotes the surface S from which the contribution of the singularity at $R=0$ has been removed. Since $R\neq 0$ inside S_0, the curl operator can be then taken inside the integral. Now, substitution of (2.19) in (2.18) results in

$$\frac{\mathbf{J}(\mathbf{r},t)}{2} - \hat{\mathbf{n}}\times\frac{1}{4\pi}\int_{S_0}\nabla\times\frac{\mathbf{J}(\mathbf{r}',\tau)}{R}\mathrm{d}S' = \hat{\mathbf{n}}\times\mathbf{H}^i(\mathbf{r},t). \tag{2.20}$$

According to the vector identity

$$\nabla\times\left(\frac{\mathbf{J}}{R}\right) = \frac{1}{R}\nabla\times\mathbf{J} + \nabla\left(\frac{1}{R}\right)\times\mathbf{J}, \qquad \nabla_{\mathbf{r}}\left(\frac{1}{R}\right) = -\frac{1}{R^2}\frac{\mathbf{R}}{R} \tag{2.21}$$

where the curl of time-varying retarded current is obtained using simple algebra,

$$\nabla \times \mathbf{J}(\mathbf{r}', \tau) = \frac{1}{c}\frac{\partial}{\partial t}\mathbf{J}(\mathbf{r}', \tau) \times \frac{\mathbf{R}}{R}. \tag{2.22}$$

Therefore, the curl of the integrand in (2.20) can be represented by

$$\nabla \times \frac{\mathbf{J}(\mathbf{r}', \tau)}{R} = \frac{1}{c}\frac{\partial}{\partial t}\mathbf{J}(\mathbf{r}', \tau) \times \frac{\mathbf{R}}{R^2} + \mathbf{J}(\mathbf{r}', \tau) \times \frac{\mathbf{R}}{R^3}. \tag{2.23}$$

Finally, the combination of (2.20) and (2.23) yields in

$$\frac{\mathbf{J}(\mathbf{r}, t)}{2} - \frac{1}{4\pi}\hat{\mathbf{n}} \times \int_{S_0} \left[\frac{1}{c}\frac{\partial}{\partial t}\mathbf{J}(\mathbf{r}', \tau) \times \frac{\mathbf{R}}{R^2} + \mathbf{J}(\mathbf{r}', \tau) \times \frac{\mathbf{R}}{R^3}\right] \mathrm{d}S' = \hat{\mathbf{n}} \times \mathbf{H}^i(\mathbf{r}, t) \tag{2.24}$$

or

$$\frac{\mathbf{J}(\mathbf{r}, t)}{2}\Omega(\mathbf{r}) - \hat{\mathbf{n}} \times \frac{1}{4\pi} P\,V \int_S \left[\frac{R}{c}\frac{\partial}{\partial t} + 1\right] \mathbf{J}(\mathbf{r}', \tau) \times \frac{\mathbf{R}}{R^3}\mathrm{d}\mathbf{S}' = \hat{\mathbf{n}} \times \mathbf{H}^i(\mathbf{r}, t)$$

where $\Omega(\mathbf{r})$ stands for the solid angle (subtended at the field location) external to the surface [44]. An alternative version of the MFIE which is identically solved for the surface field $\mathbf{H}(\mathbf{r}, t)$ rather than the surface current $\hat{\mathbf{n}} \times \mathbf{H}(\mathbf{r}, t)$ as stated in (2.24)

$$\frac{\mathbf{H}(\mathbf{r}, t)}{2} - \frac{1}{4\pi} \int_{S_0} \left[\frac{1}{c}\frac{\partial}{\partial t}(\hat{\mathbf{n}} \times \mathbf{H}(\mathbf{r}', \tau)) \times \frac{\mathbf{R}}{R^2} + (\hat{\mathbf{n}} \times \mathbf{H}(\mathbf{r}', \tau)) \times \frac{\mathbf{R}}{R^3}\right] \mathrm{d}S' = \mathbf{H}^i(\mathbf{r}, t)$$

has been used by Walker [2] and Kawaguchi [45]. It is worth mentioning that the time domain integral form of the exterior acoustic wave equation, describing the scattering of an incident wave by a hard body,

$$\frac{p(\mathbf{r}, t)}{2} - \frac{1}{4\pi} \int_{S_0} \left[\frac{R}{c}\frac{\partial}{\partial t}p(\mathbf{r}', \tau) + p(\mathbf{r}', \tau)\right] \hat{\mathbf{n}} \cdot \frac{\mathbf{R}}{R^3}\mathrm{d}S' = p^i(\mathbf{r}, t).$$

is solved for the unknown pressure p, numerically similar to the MFIE case.

2.3 Combined Field Integral Equation (CFIE)

The CFIE is obtained through a weighted average of the EFIE and MFIE in the form of

$$\begin{aligned}(1-\kappa)\mathbf{J}(\mathbf{r}, t) - \hat{\mathbf{n}} \times \left[\hat{\mathbf{n}} \times \frac{\kappa}{\eta}\mathbf{E}^s(\mathbf{r}, t) + (1-\kappa)\mathbf{H}^s(\mathbf{r}, t)\right] \\ = \hat{\mathbf{n}} \times \left[\hat{\mathbf{n}} \times \frac{\kappa}{\eta}\mathbf{E}^i(\mathbf{r}, t) + (1-\kappa)\mathbf{H}^i(\mathbf{r}, t)\right]\end{aligned}$$

where $\kappa \in \mathbb{R}$ is a dimensionless parameter in the range of $0 \leq \kappa \leq 1$. As with the frequency-domain CFIE, $0 \leq \kappa \leq 1$ results in a resonance free solution [46]. The values $\kappa = 0$ and $\kappa = 1$ specify the MFIE and EFIE where only in the latter case the equation can be applied to open structures. The final moment matrix equation for the CFIE can be obtained by linear combination of the coefficient matrices and excitation vectors

$$\begin{aligned}[\alpha_{mk}][c_{k,n}] &= [\gamma_{m,n}], \qquad n = 0, 1, 2, \ldots \\ \alpha_{mk} &= \frac{\kappa}{\eta}\alpha^E_{mk} + (1-\kappa)\alpha^H_{mk} \\ \gamma_{m,n} &= \frac{\kappa}{\eta}\gamma^E_{m,n} + (1-\kappa)\gamma^H_{m,n}\end{aligned} \tag{2.25}$$

pertinent to the EFIE and MFIE cases.

2.3.1 Dielectric Scatterer

Consider a homogeneous dielectric body with a permittivity ϵ_2 and a permeability μ_2 placed in an infinite homogeneous medium with a permittivity ϵ_1 and a permeability μ_1. By invoking the equivalence principle, the scattered field can be formulated in terms of the equivalent electric and magnetic current $\mathbf{J}$ and $\mathbf{M}$ on the surface S of the dielectric body. After enforcing the continuity of the tangential electric field at S^- and S^+, the following integral equation is obtained to find the induced equivalent currents on the surface of dielectric object:

$$\hat{\mathbf{n}} \times \left(\hat{\mathbf{n}} \times \mathbf{E}^s_\nu(\mathbf{r},t)\right) = \begin{cases} -\hat{\mathbf{n}} \times (\hat{\mathbf{n}} \times \mathbf{E}^i(\mathbf{r},t)), & \nu = 1 \qquad \forall \mathbf{r} \in S^- \\ 0, & \nu = 2 \qquad \forall \mathbf{r} \in S^+ \end{cases}$$

where ν represents the medium in which the scattered field is evaluated. When the equivalent currents are placed, because in case $\nu = 1$ the field in the interior region to the dielectric body is zero, the entire space can be assumed to be filled with the material (ϵ_1,μ_1), and vice versa for the case $\nu = 2$, since the field in the exterior region to the body is zero, the external region can be replaced by the material (ϵ_2,μ_2). As a result, the scattered electric field due to the equivalent electric and magnetic currents $\mathbf{J}$ and $\mathbf{M}$ is given by

$$\mathbf{E}^s_\nu(\mathbf{r},t) = -\frac{\partial \mathbf{A}_\nu(\mathbf{r},t)}{\partial t} - \nabla\phi_\nu(\mathbf{r},t) - \frac{1}{\epsilon_\nu}\nabla \times \mathbf{F}_\nu(\mathbf{r},t) \tag{2.26}$$

where $\mathbf{F}_\nu$ is the electric vector potential in medium ν,

$$\mathbf{F}_\nu(\mathbf{r},t) = \frac{\epsilon_\nu}{4\pi}\int_S \frac{\mathbf{M}(\mathbf{r}',\tau_\nu)}{R}\mathrm{d}S', \tag{2.27}$$

and the definitions of $\mathbf{A}_\nu$ and ϕ_ν are identical to (2.4) and (2.6),

$$\mathbf{A}_\nu(\mathbf{r},t) = \frac{\mu_\nu}{4\pi}\int_S \frac{\mathbf{J}(\mathbf{r}',\tau_\nu)}{R}\mathrm{d}S'$$

$$\phi_\nu(\mathbf{r},t) = \frac{1}{4\pi\epsilon_\nu}\int_S \frac{\sigma(\mathbf{r}',\tau_\nu)}{R}\mathrm{d}S',$$

in which the retarded time $\tau_\nu = t - \frac{R}{c_\nu}$ and the velocity of propagation of the EM wave in the medium (ϵ_ν,μ_ν) is $c_\nu = \frac{1}{\sqrt{\epsilon_\nu\mu_\nu}}$. Combining (2.26) and the enforced continuity condition on the electric field gives the EFIE

$$\left[\frac{\partial \mathbf{A}_\nu(\mathbf{r},t)}{\partial t} + \nabla\phi_\nu(\mathbf{r},t) + \frac{1}{\epsilon_\nu}\nabla \times \mathbf{F}_\nu(\mathbf{r},t)\right]_{\tan} = \begin{cases} \mathbf{E}^i(\mathbf{r},t), & \nu = 1 \\ 0, & \nu = 2 \end{cases} \qquad \forall \mathbf{r} \in S. \tag{2.28}$$

Dual to the EFIE formulation, one can obtain the MFIE formulation.

2.3.2 Narrow-Band Formulations

The size of the time step needed for the numerical solution of the mentioned TDIE formulations is inversely proportional to the highest frequency component of the incident transient field. This causes inefficiency when the incident pulse is narrow band in nature.

The narrow band formulation of the TDIE [47] instead demands time step sizes inversely proportional to the bandwidth of the incident signal. According to the complex envelope definition in Appendix (8.1), the temporally bandlimited version of the TDIE are obtained by substituting

$$\mathbf{J}(\mathbf{r},t) = \mathrm{Re}\left[\mathcal{J}(\mathbf{r},t)e^{j\omega_o t}\right] \quad (2.29)$$

$$\mathbf{H}^i(\mathbf{r},t) = \mathrm{Re}\left[\mathcal{H}^i(\mathbf{r},t)e^{j\omega_o t}\right] \quad (2.30)$$

$$\mathbf{E}^i(\mathbf{r},t) = \mathrm{Re}\left[\mathcal{E}^i(\mathbf{r},t)e^{j\omega_o t}\right]$$

for the current and field values respectively in DEFIE (2.11) and MFIE (2.24),

$$\frac{\mu}{4\pi}\int_S \left[\frac{\partial^2 \mathcal{J}(\mathbf{r}',\tau)}{\partial t^2} + 2j\omega_0 \frac{\partial \mathcal{J}(\mathbf{r}',\tau)}{\partial t} - \omega_0^2 \mathcal{J}(\mathbf{r}',\tau)\right] \frac{e^{-jk_0 t}}{R} \mathrm{d}S' - \frac{\nabla_{\mathbf{r}}}{4\pi\epsilon}\int_S \nabla_{\mathbf{r}'}.\mathcal{J}(\mathbf{r}',\tau)\frac{e^{-jk_0 t}}{R}\mathrm{d}S' = \frac{\partial \underline{\mathcal{E}}^i(\mathbf{r},t)}{\partial t} + j\omega_0 \underline{\mathcal{E}}^i(\mathbf{r},t) \quad (2.31)$$

$$\frac{\mathcal{J}(\mathbf{r},t)}{2} - \frac{1}{4\pi}\hat{\mathbf{n}} \times \int_{S_0} \frac{1}{c}\left[\frac{\partial \mathcal{J}(\mathbf{r}',\tau)}{\partial t} + j\omega_0 \mathcal{J}(\mathbf{r}',\tau)\right] \times \mathbf{R}\frac{e^{-jk_0 t}}{R^2} + \mathcal{J}(\mathbf{r}',\tau) \times \mathbf{R}\frac{e^{-jk_0 t}}{R^3}\mathrm{d}S' = \hat{\mathbf{n}} \times \mathcal{H}^i(\mathbf{r},t) \quad (2.32)$$

where $k_0 = \frac{\omega_0}{c}$ is the wave number at the center frequency ω_0. If $\omega_0 = 0$, the equations gives the standard TDIE (2.11) and (2.24). When the complex envelopes of the representing current and fields $\mathcal{J}$, $\mathcal{E}$, and $\mathcal{H}$ are assumed constant, i.e. $\omega_0 \neq 0$, $\mathbf{J}$, $\mathbf{E}$, and $\mathbf{H}$ are time-harmonic and the frequency-domain integral equations are derived.

2.4 Spatial Discretization Using Vector Basis Functions (BF)

To numerically solve the TDIEs based on the moment method, the induced surface current density $\mathbf{J}(\mathbf{r},t)$ is expanded in terms of the set of vector spatial basis functions weighted by time-varying unknown coefficients

$$\mathbf{J}(\mathbf{r},t) = \sum_{k=1}^{M} I_k(t)\mathbf{f}_k(\mathbf{r}). \tag{2.33}$$

Substitution of the expansion (2.33) in the EFIE (2.10), the DEFIE (2.11) or the MFIE (2.24), respectively, gives

$$\frac{\mu}{4\pi}\sum_{k=1}^{M}\int_S \frac{\partial I_k(\tau)}{\partial \tau}\frac{\mathbf{f}_k(\mathbf{r}')}{R}\mathrm{d}S' - \frac{\nabla_{\mathbf{r}}}{4\pi\epsilon}\sum_{k=1}^{M}\int_S\int_0^{\tau} I_k(t')\frac{\nabla_{\mathbf{r}'}.\mathbf{f}_k(\mathbf{r}')}{R}\mathrm{d}t'\mathrm{d}S' = \underline{\mathbf{E}}^i(\mathbf{r},t) \tag{2.34}$$

$$\frac{\mu}{4\pi}\sum_{k=1}^{M}\int_S \frac{\partial^2 I_k(\tau)}{\partial \tau^2}\frac{\mathbf{f}_k(\mathbf{r}')}{R}\mathrm{d}S' - \frac{\nabla_{\mathbf{r}}}{4\pi\epsilon}\sum_{k=1}^{M}\int_S I_k(\tau)\frac{\nabla_{\mathbf{r}'}.\mathbf{f}_k(\mathbf{r}')}{R}\mathrm{d}S' = \frac{\partial \underline{\mathbf{E}}^i(\mathbf{r},t)}{\partial t} \tag{2.35}$$

$$\frac{1}{2}\sum_{k=1}^{M} I_k(t)\mathbf{f}_k(\mathbf{r}') - \frac{1}{4\pi}\hat{\mathbf{n}}\times\left[\sum_{k=1}^{M}\int_S \frac{\partial I_k(\tau)}{c\partial\tau}\frac{\mathbf{f}_k(\mathbf{r}')\times\mathbf{R}}{R^2}\mathrm{d}S'\right.$$
$$\left.+\sum_{k=1}^{M}\int_S I_k(\tau)\frac{\mathbf{f}_k(\mathbf{r}')\times\mathbf{R}}{R^3}\mathrm{d}S'\right] = \hat{\mathbf{n}}\times\mathbf{H}^i(\mathbf{r},t). \tag{2.36}$$

To remove the cumbersome integral on space-time, it is assumed that the current does not vary appreciably with time within the subdomains. As a result, one can extract the τ-dependent terms from the surface integrals in equations (2.34)-(2.36) and separate the temporal variation from the space quadratures.

2.4.1 Divergence-Conforming Rao-Wilton-Glisson (RWG) BF

The structure to be analyzed is first approximated by planar triangular patches which have the ability to conform to any geometrical surface of the boundary. The spatial variation of the current is then modeled by divergence-conforming vector basis functions with constant tangential and linear normal distribution, also known as triangular "roof-top" or Rao-Wilton-Glisson (RWG) vector basis functions [48]. They are defined for each common edge joining two flat triangular panels $T_n^{\pm}$ by means of

$$\mathbf{f}_m(\mathbf{r}) = \mathbf{f}_m^{+}(\mathbf{r}) + \mathbf{f}_m^{-}(\mathbf{r}), \qquad \mathbf{f}_m^{\pm}(\mathbf{r}) = \begin{cases} \frac{l_m}{2A_m^{\pm}}\rho_m^{\pm}, & \mathbf{r}\in T_m^{\pm} \\ 0, & \mathbf{r}\notin T_m^{\pm} \end{cases} \tag{2.37}$$

where the l_m reflects the length of the common edge and $A_m^{\pm}$ represents the area of the triangle $T_m^{\pm}$. According to Fig. 2.2, the variable $\rho_m^{\pm}$ is defined as the position vector with respect to the free vertex of $T_m^{\pm}$. The current represented in terms of (2.37) has no normal component (expressing line charges) along the outer boundary of the triangle pair. The current normal to the common edge is constant and continuous across the edge. Therefore, all the edges of the triangles are free of line charges. The surface divergence of the RWG basis functions is constant on each triangle,

$$\nabla_s \cdot \mathbf{f}_m^{\pm}(\mathbf{r}) = \begin{cases} \pm\frac{l_m}{A_m^{\pm}}, & \mathbf{r} \in T_m^{\pm} \\ 0, & \mathbf{r} \notin T_m^{\pm} \end{cases}. \tag{2.38}$$

The surface divergence (2.38) is proportional to the surface charge density. As a result, the total charge density associated with each triangle pair is zero.

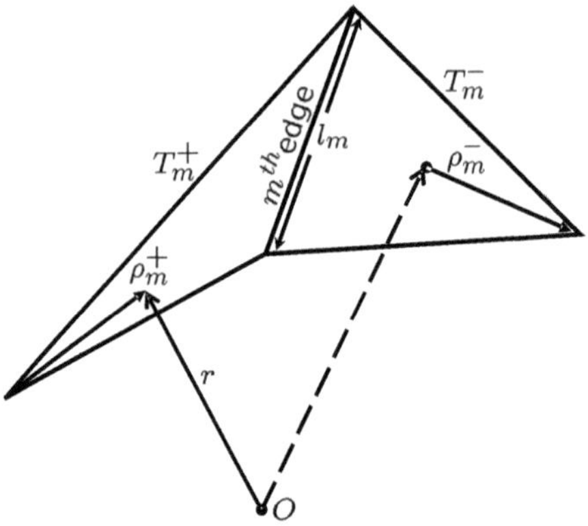

Figure 2.2: The RWG spatial vector basis function: a triangle pair with geometrical parameters, the global and local coordinates associated with the mth common edge.

The flat-faced approximation of curved scatterers causes that the normal vector assigned to every triangular patch deviates from the local normal direction of the original surface at the observation point, and hence, degrades the accuracy of the common barycentric point approximation of the Galerkin testing procedure (Section 2.5) in the boundary element method. To further improve the accuracy and computational efficiency of the integral solvers, the cylindrical parts of the object are supplanted by structured rectangular meshes whose unit normal vectors (at their centroid) coincide with the radial direction of the underlying cylindrical coordinate system.

2.4.2 Roof-Top (RT) BF

As mentioned earlier, the initial phase of the surface integral equation (SIE) solvers describing the EM radiation phenomena is the expansion of the induced surface current in terms of a set of subdomain (temporal and) spatial basis functions weighted by corresponding unknown coefficients. As the BF in the frequency-domain method of moment (MoM) [49], the RWG functions defined on triangular subdomains (Section 2.4.1) and roof-top (RT) functions on quadrilateral (rectangular) subdomains [50] are most commonly used. The RWG BF are utilized for flat-paved approximations of arbitrary shapes whereas the RT BF are conspicuous for modeling geometries that conform to Cartesian coordinates [51], [52]. In 3D EM scattering problems, generally the RWG BF are solely applied to discretize the SIEs. Although, the RT BF can be substituted on interior flat regions to reduce the number of unknowns [51], [52], the RWG BF are kept to improve the approximation of irregular boundaries [53]. In commercial software, the mixture of quadrilateral [50] and rectangular [54] with triangular cells (to fit the irregular boundaries [54]) are utilized to solve scattering problems in frequency-domain efficiently, e.g., the RT BF accommodate modeling of metallic thickness of the microstrip lines, lead frames, etc. [54]. Commercial software use the so-called double-node technique for automatic segmentation [50]. The resulting flat quadrilaterals (with triangles in between) are suitable for modeling polygonal plates.

The radiation analysis of geometrically complicated structures involving cylindrical parts are of great practical interest, e.g., simulating accelerator cavities excited by travelling charges (Section 6.11), circular waveguides (Section 6.11.2), optical resonators, EM compatibility of cable-loaded aircrafts, cylindrical reflectors, via interconnects, coaxial connectors, etc. In the vast majority of such internal scattering applications, the radial field excites the structure in a presence of axially cylindrical conductive shells. On the cylindrical parts of the scatterer, however, the normal vector over triangle subdomains, particularly at the centroid of the surface patches, is not codirectional with radial unit vector of local cylindrical coordinate systems that the tube shells stand on. The destructive effect of these mismatching orientations is aggravated in approximating long smooth cylindrical bodies by the pineapple-like coverings produced through triangular meshing of curvatures. For proper alignment of the surface normal vector, the cylindrical parts of the scatterer are covered by the RT BF Fig. 2.3. The RT BF define as

$$\mathbf{f}_m^{\pm}(\mathbf{r}) = \begin{cases} \frac{l_m}{A_m^{\pm}}(\rho_m^{\pm} \cdot \hat{u}^{\pm})\hat{u}^{\pm}, & \mathbf{r} \in P_m^{\pm} \\ 0, & \mathbf{r} \notin P_m^{\pm} \end{cases} \tag{2.39}$$

The surface divergence of (2.39) is the same as (2.38).

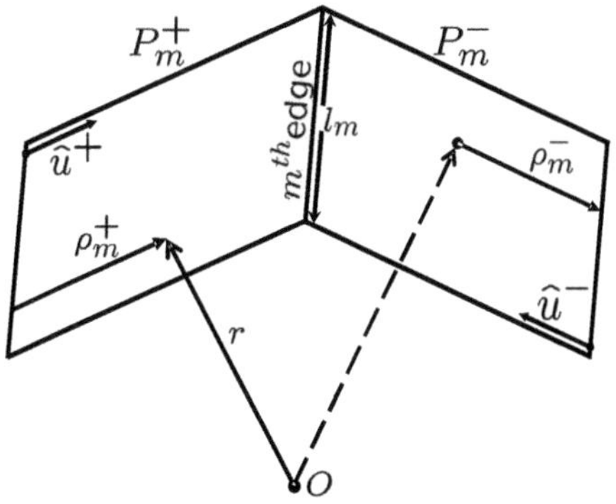

Figure 2.3: The RT spatial vector basis function: a parallelogram pair with geometrical parameters, the global and local coordinates associated with the mth common edge.

2.4.3 Linearly-Varying Hybrid BF

The flat quadrilaterals in Section 2.4.2 can not be defined by four arbitrary points in the space and obviously such elements are not suitable for modeling curved surfaces. Hence, the well-known triangles are widely used for describing arbitrary curved objects. Let the structure be meshed by a set of planar subdomains S_q involving either triangle $T_{q'}$ or rectangles $P_{q''}$, where the subscripts are all integers. This combination forms doublets over every two connected neighboring + and − subdomains, totally M unique hinging edges. Subsequently, $\mathbf{J}(\mathbf{r})$ is approximated by a merged set of subdomain vector BF,

$$\mathbf{f}_m(\mathbf{r}) = \mathbf{f}_m^+(\mathbf{r}) + \mathbf{f}_m^-(\mathbf{r}), \quad \mathbf{f}_m^\pm(\mathbf{r}) = \begin{cases} \frac{l_m}{2A_m^\pm}\rho_m^\pm, & \mathbf{r} \in T_m^\pm \\ \frac{l_m}{A_m^\pm}(\rho_m^\pm \cdot \hat{u}_m^\pm)\hat{u}_m^\pm, & \mathbf{r} \in P_m^\pm, \quad m = 1, \ldots, M \end{cases} \tag{2.40}$$

where the parameter l_m denotes the length of the common edge defined by the two shared nodes between the assumed + and − patches, and $A_m^\pm$ represents the area of the associated subdomain S_q. According to Fig. 2.4(a) and Fig. 2.4(b), the variable $\rho_m^\pm$ is defined as the position vector with respect to the free vertex of the corresponding S_q. When S_q refers to a rectangle, $\rho_m^{+(-)}$ can be defined from (to) any of the free vertices and $u_m^\pm$ indicates the unit vector on the + or − patch plane, normal to the edge (when S_q plays the role of + side, u_m is pointing to and for the − one escaping from the edge). As the definition (2.40) expresses, the RWG over $T_m^\pm$, RT on $P_m^\pm$, and mated RWG-RT on triangle-rectangle pairs (as shown in Fig. 2.4) all are zeroth-order divergence-conforming vector BF with constant normal and piecewise linear tangential components (continuous distribution) with no existence of line charges across the common interfaces of adjacent BF. The (time-varying) unknown coefficients (2.37) are assigned to each common edge joining two touching subdomains where the direction of the current is either toward (+) or outward (-) to the common edge at every point on the joint subdomains. To bypass all boundary edges of open structures, a zero value is specified for the individual open sides of marginal S_q. Fictitious zero-valued forth edges are supposed for triangles so as to unified all the computational routines regardless of the number of subdomain S_q sides. Eventually, the solution procedure skips the zero-indexed null edges.

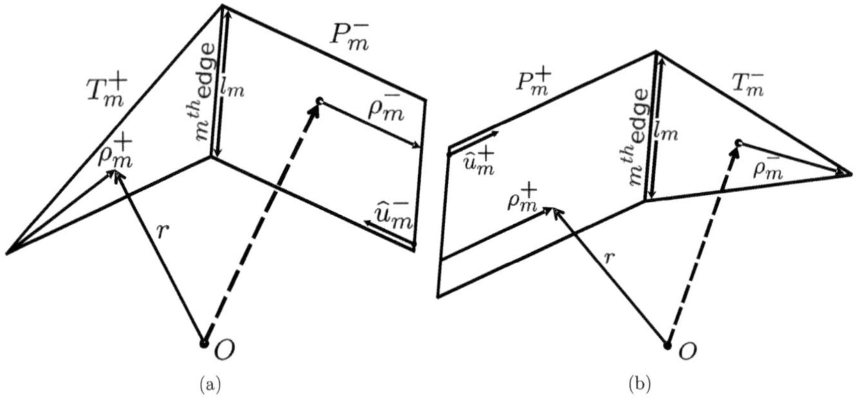

Figure 2.4: Mixed RWG-RT BF typical cases; The continuity equation along the junction of neighboring patches is automatically satisfied and the current component normal to the remaining edges is zero.

2.4.4 Mesh Plantation along Generatrices

For 3D automatic triangular mesh generation, optimization, and hierarchical refinement the open-source tool NETGEN can be utilized. Accelerator cavities, however, are mainly composed of circular, stepped, ridged, or tapered waveguides with long cylindrical sections that are excited by a bunch of moving charges, Section 6.11. In such circumstances, the moving excitation source generates a time-varying radial field $E^i(z,t)\hat{a}_r$ [55]. To excite and test the SIE properly, the value of the field at the barycenter of the observation subdomains is needed to be calculated precisely. However, as Fig. 2.5 illustrates, the normal vector at the center of triangles $\hat{n}'_r$ strays from the actual $\hat{a}_r$ which is not the case for the normal at the middle of rectangles $\hat{n}_r$. This drawback also exists when the higher-order geometrical modeling of curved triangular patches is applied. It also can not be remedied pricelessly by higher-order RWG BF transformed from the original mesh to the parametric space of a curved surface [52]. The normal vectors, but of course, are rectified when cylindrical patches are applied. For the MFIE, $\hat{n}_r$ appears besides the excitation term in the core of the functional operator (2.24). The following paragraph explains how the rectangular patches are settled upon the generatrices (mainly the cylindrical parts). Generatrices here means the bodies of translation (and revolution), e.g., the trace of a circular generatrix forms a cylindrical tube.

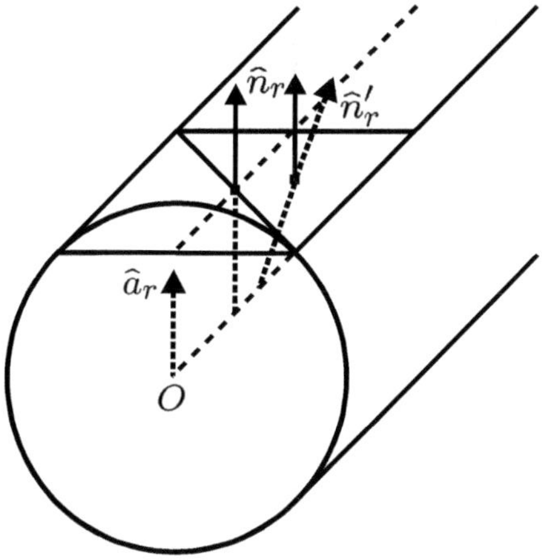

Figure 2.5: Deflection of the local normal vector passing through the centroid of the triangle cell.

The efficiency of the equilateral patches are optimal when they have regular shape as rectangular (as close as possible to square) and the numerical integrations are facilitated when the sides are parallel [50]. Hence, the radius of the circular generatrices is better not to be changed so as to prevent creation of isosceles trapezoid subdomains over the intermediate spindles or funnel-shaped parts. The resulting rectangle-triangle pairs have to be properly matched so as to satisfy the continuity equation at the interconnects. Therefore, in order to generate hybrid meshes for 3D geometries in the desired manner, first a premodel of the structure is drawn by excluding the trace path of generatrices via cutting planes and combining the irregular parts. The premodel is entirely meshed in a surface triangular mesh generator software, e.g. this can be well performed by CST MICROWAVE STUDIO® (MWS) [56]. The node coordinates and point numberings are then imported into the working environment. Knowing the longitudinal coordinate of the constant cutting planes z_L and z_R, the rectangular meshes are then planted on the absent parts by translation of the leading cutting edges. The coordinates of the rotational cutting nodes are directly applied for revolution. The edges only on the cutting plane (commonly a constant cut) needs to be renumbered for the transferred parts. The transposal size of rectangular patches is assigned by the side length of edge-triangles and the longitudinal length of them can be adjusted proportionally. See Figures in Section 6.5. Refining the triangular mesh of the premodel, the algorithm renders proportionally smaller rectangular grids. As a result, the final mesh quality is adjusted by the premodel. This technique also allows the generatrix to be transferred on a longitudinally non-uniformly distributed set of nodes whereby the smaller edges can be manually adjusted close to the wedged generatrix ends.

Other sorts of spatial basis functions can also be integrated with slight modifications to all schemes presented in the coming chapters, such as Trintinalia-Ling (linear-linear) BF [57, 58, 59], the curl-conforming BF [60], or the paired pulse BF [61, 62]. For the surface−wire junction treatment one can assume, a linear (hat) basis function $\mathbf{f}_m^w(\mathbf{r})$ on every pair of connected wire segments (S_m^- and S_m^+), that can be expressed as

$$\mathbf{f}_m^w(\mathbf{r}) = \mathbf{f}_m^{w+}(\mathbf{r}) + \mathbf{f}_m^{w-}(\mathbf{r}), \qquad \mathbf{f}_m^{\pm}(\mathbf{r}) = \begin{cases} \pm\frac{|\mathbf{r}_{m\pm1}-\mathbf{r}|}{s_m^\pm}\hat{s}_m^\pm, & \mathbf{r} \in S_m^\pm \\ 0, & \mathbf{r} \notin S_m^\pm \end{cases} \tag{2.41}$$

where $s_m^\pm = |\mathbf{r}_{m\pm1} - \mathbf{r}_m|$ and $\hat{s}_m^\pm = \frac{\mathbf{r}_{m\pm1}-\mathbf{r}_m}{s_m^\pm}$ are the unit vector tangent to segments $S_m^\pm$. The wire feeds may excite the structure by the magnetic Frill model [63].

2.5 Galerkin's Testing Procedure in Boundary Element Method

The discretized EFIE versions (2.34) and (2.35) are spatially tested in the Galerkin context, i.e. using the same weighting functions as the expansion vector functions $\mathbf{f}_m(\mathbf{r})$, $m = 1, 2, \ldots, M$, respectively, resulting in

$$\frac{\mu}{4\pi}\sum_{k=1}^{M}\frac{\partial I_k(\tau)}{\partial\tau}\int_S \mathbf{f}_m(\mathbf{r})\cdot\int_S\frac{\mathbf{f}_k(\mathbf{r}')}{R}\mathrm{d}S'\mathrm{d}S$$
$$+\frac{1}{4\pi\epsilon}\sum_{k=1}^{M}\int_0^{\tau} I_k(t')\mathrm{d}t'\int_S \nabla_{\mathbf{r}}\cdot\mathbf{f}_m(\mathbf{r})\int_S\frac{\nabla_{\mathbf{r}'}\cdot\mathbf{f}_k(\mathbf{r}')}{R}\mathrm{d}S'\mathrm{d}S = \int_S \mathbf{f}_m(\mathbf{r})\cdot\underline{\mathbf{E}}^i(\mathbf{r},t)\mathrm{d}S \tag{2.42}$$

and

$$\frac{\mu}{4\pi}\sum_{k=1}^{M}\frac{\partial^2 I_k(\tau)}{\partial\tau^2}\int_S \mathbf{f}_m(\mathbf{r})\cdot\int_S\frac{\mathbf{f}_k(\mathbf{r}')}{R}\mathrm{d}S'\mathrm{d}S$$
$$+\frac{1}{4\pi\epsilon}\sum_{k=1}^{M} I_k(\tau)\int_S \nabla_{\mathbf{r}}\cdot\mathbf{f}_m(\mathbf{r})\int_S\frac{\nabla_{\mathbf{r}'}\cdot\mathbf{f}_k(\mathbf{r}')}{R}\mathrm{d}S'\mathrm{d}S = \int_S \mathbf{f}_m(\mathbf{r})\cdot\frac{\partial\underline{\mathbf{E}}^i(\mathbf{r},t)}{\partial t}\mathrm{d}S \tag{2.43}$$

in which the vector identity $\nabla_{\mathbf{r}} \cdot (\phi \mathbf{f}_m) = \mathbf{f}_m \cdot \nabla_{\mathbf{r}}\phi + \phi \nabla_{\mathbf{r}} \cdot \mathbf{f}_m$ in conjunction with the properties of $\mathbf{f}_m$ have been used in testing the gradient of the scalar potential $\nabla_{\mathbf{r}}\phi$ in (2.34) or the gradient of the Hertz potential $\nabla_{\mathbf{r}}\Phi$ in (2.35). Similarly, taking the inner product of the testing functions $< \mathbf{f}_m(\mathbf{r}), \cdot >$ with both sides of the MFIE (2.36) gives

$$\frac{1}{2}\sum_{k=1}^{M} I_k(t)\int_S \mathbf{f}_m(\mathbf{r})\cdot\mathbf{f}_k(\mathbf{r}')\mathrm{d}S - \frac{1}{4\pi}\left[\sum_{k=1}^{M}\frac{\partial I_k(\tau)}{c\partial\tau}\int_S \mathbf{f}_m(\mathbf{r})\cdot\hat{\mathbf{n}}\times\int_S \frac{\mathbf{f}_k(\mathbf{r}')\times\mathbf{R}}{R^2}\mathrm{d}S'\mathrm{d}S \right.$$
$$\left. +\sum_{k=1}^{M} I_k(\tau)\int_S \mathbf{f}_m(\mathbf{r})\cdot\hat{\mathbf{n}}\times\int_S \frac{\mathbf{f}_k(\mathbf{r}')\times\mathbf{R}}{R^3}\mathrm{d}S'\mathrm{d}S\right] = \int_S \mathbf{f}_m(\mathbf{r})\cdot\hat{\mathbf{n}}\times\mathbf{H}^i(\mathbf{r},t)\mathrm{d}S. \qquad (2.44)$$

Note that some references, such as [6], show the double surface integrals by quadruple quadratures

$$\int_S\int_S \mathbf{G}(\mathbf{r},\mathbf{r}')\mathrm{d}S'\mathrm{d}S \equiv \iint_S \iint_S \mathbf{G}(\mathbf{r},\mathbf{r}')\mathrm{d}\mathbf{r}'\mathrm{d}\mathbf{r}. \qquad (2.45)$$

To formulate all the relevant computational schemes in a generalized framework facilitating better comparison as well as easier and more efficient implementation, the following retarded-time independent terms are considered in succession:

$$\begin{aligned}
A_{mk} &= \frac{\mu}{4\pi}\int_S \mathbf{f}_m(\mathbf{r})\cdot\int_S \frac{\mathbf{f}_k(\mathbf{r}')}{R}\mathrm{d}S'\mathrm{d}S \\
B_{mk} &= \frac{1}{4\pi\epsilon}\int_S \nabla_{\mathbf{r}}\cdot\mathbf{f}_m(\mathbf{r})\int_S \frac{\nabla_{\mathbf{r}'}\cdot\mathbf{f}_k(\mathbf{r}')}{R}\mathrm{d}S'\mathrm{d}S \\
C_{mk} &= \frac{1}{4\pi c}\int_S \mathbf{f}_m(\mathbf{r})\cdot\hat{\mathbf{n}}\times\int_S \frac{\mathbf{f}_k(\mathbf{r}')\times\mathbf{R}}{R^2}\mathrm{d}S'\mathrm{d}S \\
D_{mk} &= \frac{1}{4\pi}\int_S \mathbf{f}_m(\mathbf{r})\cdot\hat{\mathbf{n}}\times\int_S \frac{\mathbf{f}_k(\mathbf{r}')\times\mathbf{R}}{R^3}\mathrm{d}S'\mathrm{d}S \\
F_{mk} &= \frac{1}{2}\int_S \mathbf{f}_m(\mathbf{r})\cdot\mathbf{f}_k(\mathbf{r}')\mathrm{d}\mathbf{S} \\
e_m &= \int_S \mathbf{f}_m(\mathbf{r})\cdot\hat{\mathbf{n}}\times\left(\hat{\mathbf{n}}\times\mathbf{E}^i(\mathbf{r},t)\right)\mathrm{d}S \\
\dot{e}_m &= \int_S \mathbf{f}_m(\mathbf{r})\cdot\hat{\mathbf{n}}\times\left(\hat{\mathbf{n}}\times\frac{\partial\mathbf{E}^i(\mathbf{r},t)}{\partial t}\right)\mathrm{d}S \\
h_m &= \int_S \mathbf{f}_m(\mathbf{r})\cdot\hat{\mathbf{n}}\times\mathbf{H}^i(\mathbf{r},t)\mathrm{d}S.
\end{aligned}$$

The incorporation of the above terms with the definition of vector basis functions (2.37) reveals that the following integrals are needed to be evaluated once:

$$A_{mk}^{pq} = \frac{1}{A_m^p A_k^q} \int_{T_m^p} \rho_m^p \cdot \int_{T_k^q} \frac{\rho'^q_k}{R} \mathrm{d}S' \mathrm{d}S \tag{2.46}$$

$$B_{mk}^{pq} = \frac{1}{A_m^p A_k^q} \int_{T_m^p} \int_{T_k^q} \frac{1}{R} \mathrm{d}S' \mathrm{d}S \tag{2.47}$$

$$C_{mk}^{pq} = \frac{1}{A_m^p A_k^q} \int_{T_m^p} \rho_m^p \cdot \hat{\mathbf{n}} \times \int_{T_k^q} \frac{\rho'^q_k \times \mathbf{R}}{R^2} \mathrm{d}S' \mathrm{d}S \tag{2.48}$$

$$D_{mk}^{pq} = \frac{1}{A_m^p A_k^q} \int_{T_m^p} \rho_m^p \cdot \hat{\mathbf{n}} \times \int_{T_k^q} \frac{\rho'^q_k \times \mathbf{R}}{R^3} \mathrm{d}S' \mathrm{d}S \tag{2.49}$$

$$F_{mk}^{pq} = \frac{1}{A_m^p A_k^q} \int_{T_m^p} \rho_m^p \cdot \rho'^q_k \mathrm{d}S, \tag{2.50}$$

besides the excitation term(s) that have to be evaluated at every time step t_n, $n = 1, 2, \cdots, N$:

$$e_{m,n}^p = \frac{1}{A_m^p} \int_{T_m^p} \rho_m^p \cdot \underline{\mathbf{E}}^i(\mathbf{r}, t_n) \mathrm{d}S$$

$$\dot{e}_{m,n}^p = \frac{1}{A_m^p} \int_{T_m^p} \rho_m^p \cdot \frac{\partial \underline{\mathbf{E}}^i(\mathbf{r}, t_n)}{\partial t} \mathrm{d}S$$

$$h_{m,n}^p = \frac{1}{A_m^p} \int_{T_m^p} \rho_m^p \cdot \hat{\mathbf{n}} \times \mathbf{H}^i(\mathbf{r}, t_n) \mathrm{d}S$$

where p and q are $+$ or $-$ and dyadic fractions of the constant product of the edge lengths $l_m(l_k)$ have been factored out to reduce the computational complexities. The surface integrals over the observer patches T_m^p are approximated by values of the respective integrands at the centroid of the triangles, i.e., $\rho_m^p \to \rho_m^{cp}$. "Duffy's method" is used for more exact integration around singularities, see Appendix 8.2. To calculate the inner products of vector bases in (2.50), the readers are referred to the Appendix 8.3.

2.5.1 Adaptive Space Quadrature Schemes

To calculate the elements of the coefficient matrices $\bar{\bar{\mathrm{Z}}}_n$ ($n = 0, 1, \ldots, N-1$), instead of redundant direct evaluation of the mutual coupling of edge pairs individually, triangle pairs' interactions are considered. To perform numerical integration over the source patches (the interior surface integrals), the vertices of the corresponding triangle are first mapped to the points (0,0), (0,1), and (1,0) in the xy-plane and then a p-point quadrature rule is applied to the normalized triangle. Any constant multi-point quadrature rule can be considered at this stage, e.g. the 3-point, 4-point, or 7-point quadrature rules [64]. To adaptively control the precision of numerical integrations over the surface patches so as to guarantee a total quadrature error of less than an a priori set value ε, two-scale refinements of the source patches are proposed. At the first stage, if the difference between the results of for instance a 1-point and the p-point quadrature rule is greater than the prespecified error, the source triangle is partitioned preferably into four equilateral subtriangles by bisecting and connecting the middle of its three sides, as illustrated in Fig. 2.6. Every half-scaled triangle is again mapped to the vertices of the unit triangle and the sum of their contributions is checked through comparison with the result of the previous stage. Subdivision of internal subtriangles and comparison of resultant quadrature values with those of the former stage is successively continued until a predefined precision ensuring sufficiently accurate results. As a more efficient extended case, one can also perform a higher order quadrature rule in parallel, and additionally compare the results of two quadrature routines before taking any decision for further partitioning. Note that the Jacobians of the transformation at every stage are a quarter fraction of the surface area of the underlying triangle, A_k^q, and since they are finally canceled during evaluation of (2.46)-(2.49), there is no need for explicit calculation and storage.

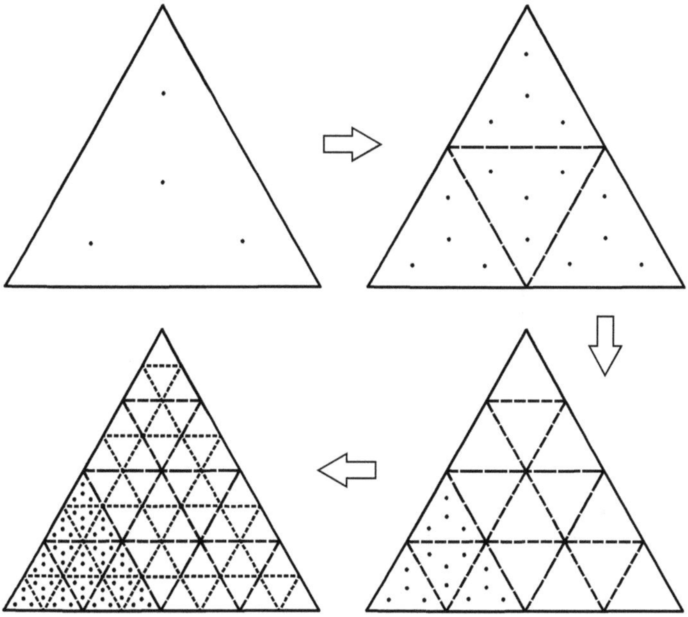

Figure 2.6: The source triangle is successively partitioned into four equal subtriangles by connecting the bisectors of every three sides, a 3-fold partitioned patch resulting in 4^3 subtriangles.

Chapter 3

Temporal Discretization

The next step after the spatial testing in Section 2.5 is the time integration methods. Generally, three kinds of time integration algorithms can be applied for the TDIE solution, namely the method of lines, the basis function expansion, and the Rothe's method. The latter solves for all time unknowns simultaneously [65] and hence it is of less interest. The former ones favorably let marching-on-in solution samples up to arbitrary time instances. Depending on the type of time discretization and whether the Galerkin's testing [21, 10] or point matching collocation method [1, 2, 3, 5, 6] in time is to be used, diverse formulas may arise for the final matrix equation construction. Accordingly, the detailed derivations of the associated final matrix equations are classified in the following Sections.

3.1 Marching-on-in-Time (MOT) Schemes

To cast the spatio-temporally discretized and tested integral equations (2.42), (2.43), and (2.44) into linear system of equations, generally, the following terms are thus needed to be computed at each time step:

$$a_{mk,n} = \frac{\mu l_m l_k}{16\pi} \sum_{p,q} a^{pq}_{mk,n} A^{pq}_{mk} \tag{3.1}$$

$$b_{mk,n} = \frac{l_m l_k}{16\pi\epsilon} \sum_{p,q} b^{pq}_{mk,n} B^{pq}_{mk} \tag{3.2}$$

$$\dot{a}_{mk,n} = \frac{\mu l_m l_k}{16\pi} \sum_{p,q} \dot{a}^{pq}_{mk,n} A^{pq}_{mk}$$

$$\dot{b}_{mk,n} = \frac{l_m l_k}{16\pi\epsilon} \sum_{p,q} \dot{b}^{pq}_{mk,n} B^{pq}_{mk}$$

$$c_{mk,n} = \frac{l_m l_k}{16\pi c} \sum_{p,q} c^{pq}_{mk,n} C^{pq}_{mk} \tag{3.3}$$

$$d_{mk,n} = \frac{l_m l_k}{16\pi} \sum_{p,q} d^{pq}_{mk,n} D^{pq}_{mk} \tag{3.4}$$

$$f_{mk,n} = \frac{l_m l_k}{8} \sum_{p,q} f^{pq}_{mk,n} F^{pq}_{mk} \tag{3.5}$$

$$e_{m,n} = \frac{l_m}{2}\sum_p e^p_{m,n} \tag{3.6}$$

$$\dot{e}_{m,n} = \frac{l_m}{2}\sum_p \dot{e}^p_{m,n}$$

$$h_{m,n} = \frac{l_m}{2}\sum_p h^p_{m,n}. \tag{3.7}$$

Note that except the excitation terms $e^p_{m,n}$ or $\dot{e}^p_{m,n}$ and/or $h^p_{m,n}$, all time-independent terms A^{pq}_{mk}, B^{pq}_{mk}, C^{pq}_{mk}, D^{pq}_{mk}, and F^{pq}_{mk} do not need to be updated for every time step. The retarded time can be approximated by

$$\tau = t - \frac{R}{c} \longrightarrow \tau^{pq}_{mk,n} = t_n - \frac{R^{pq}_{mk}}{c} \tag{3.8}$$

where $R^{pq}_{mk} = |r^{cp}_m - r'^{cq}_k|$ is the distance between the centroids of triangle T^p_m and T^q_k and depending on the temporal discretization scheme, the following retarded time-dependent scaling factors thus have to be updated at each time step, either

$$a^{pq}_{mk,n} = \frac{\partial I_k(\tau^{pq}_{mk,n})}{\partial \tau} \tag{3.9}$$

$$b^{pq}_{mk,n} = \int_0^{\tau^{pq}_{mk,n}} I_k(t)\mathrm{d}t \tag{3.10}$$

for the EFIE or

$$\dot{a}^{pq}_{mk,n} = \frac{\partial^2 I_k(\tau^{pq}_{mk,n})}{\partial \tau^2} \tag{3.11}$$

$$\dot{b}^{pq}_{mk,n} = I_k(\tau^{pq}_{mk,n}) \tag{3.12}$$

for the DEFIE, as well as

$$c^{pq}_{mk,n} = \frac{\partial I_k(\tau^{pq}_{mk,n})}{\partial \tau} \tag{3.13}$$

$$d^{pq}_{mk,n} = I_k(\tau^{pq}_{mk,n}) \tag{3.14}$$

$$f^{pq}_{mk,n} = I_k(t_n) \tag{3.15}$$

for the MFIE. Basically, for $m = 1, 2, \ldots, M$ at every time instant $t = t_n$ only the time-varying coefficient matrix $a_{mk,n}$, $b_{mk,n}$, and $e^{(')}_{m,n}$ are evaluated so as to constitute the linear system of equations

$$\sum_{k=1}^{M} (a_{mk,n} + b_{mk,n}) = e_{m,n} \tag{3.16}$$

$$\sum_{k=1}^{M} \left(\dot{a}_{mk,n} + \dot{b}_{mk,n}\right) = \dot{e}_{m,n}$$

for the EFIE versions, and $c_{mk,n}$, $d_{mk,n}$, $f_{mk,n}$, and $h_{m,n}$ are updated to cast the final algebraic system of equations

$$\sum_{k=1}^{M} (f_{mk,n} - c_{mk,n} - d_{mk,n}) = h_{m,n} \tag{3.17}$$

for the MFIE. Note that $g_{mk,n}$ where $g \in \{a, b, \dot{a}, \dot{b}, c, d, f\}$ denotes that the indices of the source and observation patches k and m respectively vary first at the nth time instance, i.e., all the spatial interactions are first considered for the fix temporal value t_n. That is, the linear system of equations (3.16) and (3.17) both have the form

$$\sum_{r=0}^{n} \bar{\bar{Z}}_{n-r}\bar{I}_r = \bar{V}_n \qquad n = 1, 2, \ldots, N \tag{3.18}$$

where $\bar{I}_0 = 0$. For solving the CFIE, (3.16) and (3.17) are linearly combined in their above form (3.18) as explicated in (2.25). The generalized distinct spatio-temporal discretization platform for implementation of the MOT schemes (3.18) helps to take advantages of the extensive research resources existing for frequency-domain BEMs in spatial discretization stage. For instance, "loop-tree" decomposition of divergence-conforming vector basis functions [66] or the higher order extensions of the vector basis functions [67, 68] or combination of them [22] or Buffa-Christiansen bases [69] can be readily incorporated into the time-domain BEM formulations, as explained in [30].

At last, those many discretized terms for which $R \geq c\Delta t$, are moved to the right-hand side of (3.16) and/or (3.17) and the unknown terms associated with the present time $t = t_n$ are retained on the left-hand side to establish the matrix equation

$$\bar{\bar{Z}}_0\bar{I}_n = \bar{V}_n - \sum_{r=1}^{n-1} \bar{\bar{Z}}_{n-r}\bar{I}_r \qquad n = 1, 2, \ldots, N \tag{3.19}$$

where the elements of the matrix $\{\bar{\bar{Z}}_0\}_{m,k}$ are not functions of time but time step size. Hence, they need to be computed only once at the first time step and preferably LU decomposed on demand. Finally, only the right-hand side of (3.19) is sequentially updated in every time step and the corresponding matrix equation (3.19) is consecutively solved at each time sample for the unknown current coefficients at the present time by performing forward/backward substitution on the LU decomposed version of $[Z_0]_{M\times M}$. This recursive construction of (3.19) and its successive solution in time is known as the MOT procedure. In the MOT solution algorithm (3.19), the unknown current coefficients are found recursively. First, $\bar{I}_1$ at time $t_1 = \Delta t$ is found; this, in turns, permits the computation of the instantaneous scattered field $\tilde{V}_2 = \bar{\bar{Z}}_1\bar{I}_1$. This vector is then subtracted from the tested incident field $\bar{V}_2$ and (3.19) is solved for the current weights $\bar{I}_2$ at time $t_2 = 2\Delta t$. At the next step, the past current samples $\bar{I}_1$ and $\bar{I}_2$ are used to compute $\tilde{V}_3 = \bar{\bar{Z}}_2\bar{I}_1 + \bar{\bar{Z}}_1\bar{I}_2$, which together with $\bar{V}_3$ permits the computation of $\bar{I}_3$, and so forth.

In time-marching methods, since the Green's function is impulsive in time, when the time basis functions are local only the sparse matrices $\bar{\bar{Z}}_0$ through $\bar{\bar{Z}}_{N_g}$ are nonzero, where

$$N_g = \left\lfloor \frac{D}{c} + \frac{\Delta T}{\Delta t} \right\rfloor, \tag{3.20}$$

in which ΔT is the duration of interpolatory temporal basis function, D is the maximum linear dimension of the scatterer (i.e., $\max(R)$, the maximum distance between any two points on S) and $\lfloor x \rfloor$ denotes the largest integer less than or equal to x. In fact, $N_g\Delta t$ approximates the longest possible transit time of the field produced by the temporal basis function $T(t)$ across S. Thus, (3.19) is recasted into

$$\bar{\bar{Z}}_0\bar{I}_n = \bar{V}_n - \sum_{r=1}^{\min(n-1,N_g)} \bar{\bar{Z}}_r\bar{I}_{n-r} \qquad n = 1, 2, \ldots, N. \tag{3.21}$$

Fig. (3.1) visualizes N_g matrix-vector products needed to be calculated for the construction of the right hand side of (3.21) at every time step. The more physically meaningful alternative representation for (3.19) is obtained once the summation terms in (3.21) are flipped as

$$\bar{\bar{Z}}_0\bar{I}_n = \bar{V}_n - \sum_{r=\max(1,n-N_g)}^{n-1} \bar{\bar{Z}}_{n-r}\bar{I}_r = \bar{V}_n - \tilde{V}_n \qquad n = 1, 2, \ldots, N. \tag{3.22}$$

It follows that at every time step the currents on the scatterer radiate fields that interact with the scatterer for at most N_g time steps before leaving the scatterer.

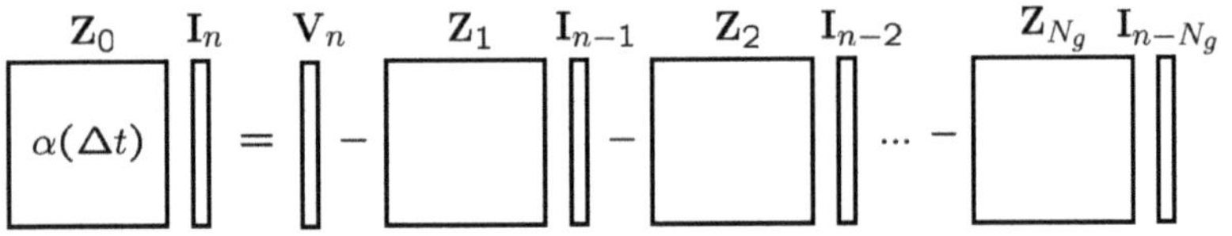

Figure 3.1: The recursive usage of recent solution samples to build the retarded fields $\tilde{V}_n$.

Now, the promising time discretization choices for stable BEMs are individually discussed in the following Sections.

3.2 Time Integration Methods

As the next stage in the numerical solution procedure of the TDIEs in (2.42)-(2.44), the time axis is divided into equal intervals Δt defining time instants $t_n = n\Delta t$.

3.2.1 Theta Method

To approximate the time derivatives, e.g. (2.9), the simplest general scheme is the theta (weighted) method

$$\frac{\mathbf{A}(\mathbf{r}, t_n) - \mathbf{A}(\mathbf{r}, t_{n-1})}{\Delta t} + \theta\nabla\phi(\mathbf{r}, t_n) + (1-\theta)\nabla\phi(\mathbf{r}, t_{n-1}) = \underline{\mathbf{E}}^i(\mathbf{r}, t_{n-(1-\theta)}) \tag{3.23}$$

where specific choices of the parameter θ lead to well-known classical methods. Namely $\theta = 0$ reflects the explicit (forward) Euler method, whereas $\theta = 1$ implies the implicit (backward) Euler method. Besides backward and forward Euler, another first-order difference method frequently used in numerical solution of differential algebraic system of equations, known as the Galerkin method, is deduced by $\theta = \frac{2}{3}$. Setting $\theta = \frac{1}{2}$ results in the only second order approximation represented by (3.23), the so-called Crank-Nicolson or implicit midpoint (trapezoidal) method, in which the time derivative associated with the vector potential term is approximated by the central finite difference and the time averaging is used for the scalar potential term. In fact, (3.23) combines all recent cases into a unified equation so that one can readily develop a unified code for implementing various time integration schemes. Generally, applying the θ-method leads to

$$a^{pq}_{mk,n} = \frac{\partial I_k(\tau^{pq}_{mk,n})}{\partial \tau} = \frac{I_k(t_r) - I_k(t_{r-1})}{\Delta t} \tag{3.24}$$

where

$$t_{r-1} < \tau^{pq}_{mk,n} \le t_r. \tag{3.25}$$

Assuming that a linear interpolation is used for approximating the values of the currents at retarded times, (3.10) is calculated for the implicit backward Euler and the explicit forward Euler respectively as

$$b^{pq}_{mk,n}|_{\theta=1} = \Delta t \sum_{t=t_0}^{t_{r-1}} I_k(t) + \left(\delta - \frac{\delta^2}{2} - \frac{1}{2}\right) I_k(t_{r-1}) + \frac{\delta^2}{2} I_k(t_r) \tag{3.26}$$

$$b^{pq}_{mk,n}|_{\theta=0} = \Delta t \sum_{t=t_0}^{t_{r-2}} I_k(t) + \left(\delta - \frac{\delta^2}{2} - \frac{1}{2}\right) I_k(t_{r-2}) + \frac{\delta^2}{2} I_k(t_{r-1}) \tag{3.27}$$

where

$$\delta = \frac{\tau^{pq}_{mk,n} - t_{r-1}}{\Delta t}. \tag{3.28}$$

Apparently, for the general case of (3.23), b^{pq}_{mk} is equal to θ times (3.26) plus $(1-\theta)$ times (3.27).

Considering the temporal variation of $\mathbf{A}(\mathbf{r}, t)$ as a quadratic polynomial, a more accurate representation may be provided by the second order backward finite difference formula

$$\frac{3\mathbf{A}(\mathbf{r}, t_n) - 4\mathbf{A}(\mathbf{r}, t_{n-1}) + \mathbf{A}(\mathbf{r}, t_{n-2})}{2\Delta t} + \nabla\phi(\mathbf{r}, t_n) = \underline{\mathbf{E}}^i(\mathbf{r}, t_n).$$

This three-point backward asymmetric scheme is also 2nd order accurate in time, $\mathcal{O}(\Delta t^2)$, and results in

$$a^{pq}_{mk,n} = \frac{3I_k(t_r) - 4I_k(t_{r-1}) + I_k(t_{r-2})}{2\Delta t}$$

while the same expression as (3.26) is obtained for $b^{pq}_{mk,n}$.

Considering the DEFIE (2.11), the second time derivative can be approximated by

$$\frac{\mathbf{A}(\mathbf{r}, t_n) - 2\mathbf{A}(\mathbf{r}, t_{n-1}) + \mathbf{A}(\mathbf{r}, t_{n-2})}{\Delta t^2} + \nabla\Phi(\mathbf{r}, t_n) = \frac{\partial \underline{\mathbf{E}}^i(\mathbf{r}, t_n)}{\partial t}. \tag{3.29}$$

For this case, (3.11) and (3.12), respectively, turn to

$$\dot{a}^{pq}_{mk,n} = \frac{I_k(t_r) - 2I_k(t_{r-1}) + I_k(t_{r-2})}{\Delta t^2} \tag{3.30}$$

$$\dot{b}^{pq}_{mk,n} = (1-\delta) I_k(t_{r-1}) + \delta I_k(t_r). \tag{3.31}$$

The Newmark-Beta formulation [70] on the DEFIE

$$\frac{\mathbf{A}(\mathbf{r}, t_n) - 2\mathbf{A}(\mathbf{r}, t_{n-1}) + \mathbf{A}(\mathbf{r}, t_{n-2})}{\Delta t^2} + \theta\nabla\Phi(\mathbf{r}, t_n)$$
$$+(1-2\theta)\nabla\Phi(\mathbf{r}, t_{n-1}) + \theta\nabla\Phi(\mathbf{r}, t_{n-2}) = \frac{\partial \underline{\mathbf{E}}^i(\mathbf{r}, t_{n-1})}{\partial t}, \qquad 0 \le \theta \le 1 \tag{3.32}$$

gives $\dot{a}^{pq}_{mk,n}$ identical to (3.30) and

$$\dot{b}^{pq}_{mk,n} = \theta(1-\delta)I_k(t_{r-3}) + [(1-\delta)+\theta(3\delta-2)]I_k(t_{r-2}) + [\delta+\theta(1-3\delta)]I_k(t_{r-1}) + \theta\delta I_k(t_r). \tag{3.33}$$

with judiciously chosen values of $\theta \geq 0.2$ [71].

Here, we also consider the time-domain MFIE (2.24) approximated by the implicit backward difference method,

$$\frac{\mathbf{J}(\mathbf{r},t_n)}{2} - \hat{\mathbf{n}} \times \frac{1}{4\pi}\int_{S_0}\left[\frac{\mathbf{J}(\mathbf{r}',\tau_n)-\mathbf{J}(\mathbf{r}',\tau_{n-1})}{c\Delta t}\times\frac{\mathbf{R}}{R^2} + \mathbf{J}(\mathbf{r}',\tau_n)\times\frac{\mathbf{R}}{R^3}\right]dS' = \hat{\mathbf{n}} \times \mathbf{H}^i(\mathbf{r},t_n) \tag{3.34}$$

where the retarded time samples $\tau_n = t_n - \frac{R}{c}$ are linearly interpolated. As a result, the discretization coefficients (3.13) and (3.14) are obtained as follows:

$$c^{pq}_{mk,n} = \frac{I_k(t_r) - I_k(t_{r-1})}{\Delta t} = a^{pq}_{mk,n}|_{\text{ in (3.24)}}$$
$$d^{pq}_{mk,n} = (1-\delta)I_k(t_{r-1}) + \delta I_k(t_r) = \dot{b}^{pq}_{mk,n}.$$

3.2.2 Time Interpolation Methods

In the TDIE, owing to the presence of delayed terms, knowledge of past solution is required not only at nodal points rather mostly somewhere in between. Unless otherwise stated, the triangular (hat) functions are used to represent the temporal evolution, resulting in unit weights for $t = t_r$ and linear interpolation to zero for $t = t_r \pm \Delta t$. In order to illustrate that the use of smoother interpolators does not necessarily enhance, but, on the contrary, shrink the extent of the stable region, the higher order interpolating functions ensuring a qth order approximation over temporal subdomains are employed as well. Considering successive orders of the Lagrange interpolants, the $q+1$ points interpolant is equivalent to q^{th} order piecewise polynomial expanded over $(q+1)\Delta t$ intervals, namely $[(r-q)\Delta t, r\Delta t]$. The value of the current at time instance τ_n thus depends on the one ahead (t_r) and the q earlier values of the discrete current coefficients behind t_r. Therefore, the overall effect of the shifted versions provides an qth order accurate interpolation over the generic r^{th} time interval, between the samples $t_{r-1} = (r-1)\Delta t$ and $t_r = r\Delta t$. The result is a continuous q^{th} order function, with a piecewise $(q-1)^{\text{th}}$ order derivative that is continuous at the integer multiples of Δt, except when the function to be interpolated is exactly smoother than the qth order. The present work checks the time-shifted Lagrange interpolants of orders $q = 1, 2, 3, \text{and } 4$ [64]. For the sake of completeness, the cubic spline interpolation and other alternatively proposed interpolants, namely the cosine squared function [19], the optimized exponential function with all-order continuous derivatives [19], and the non-differentiable sinusoidal dome interpolant [23] are also employed to describe time evolution over the temporal subdomains. However, no explicit averaging technique is used here to filter out the intrinsic high frequency oscillations of the results as posed by many previous works [1, 18, 7].

3.2.3 Delay Differential Equation (DDE) Context

Retarded potential integral equations, namely the time-domain EFIE (2.10) and MFIE (2.24), can be stated as a general form of a DDE problem,

$$\begin{cases} \frac{\partial}{\partial t}y(t) = f(t, y(t), y(t-\tau(t, y(t)))), & 0 \leq t \\ y(t) = 0, & t \leq 0. \end{cases}$$

Contrary to the numerical ordinary differential equation (ODE) methods that furnish approximate values of the solution only at nodal points, implementation of any numerical method for the solution of DDEs, e.g. (2.10) and (2.24), requires the knowledge of the approximate solution at somewhere between many past intermediate points t_r other than the nodal points t_n. Therefore, in general, the DDE method is based on continuous extension of numerical ODE schemes. To provide step by step a continuous approximation of the solution, a posteriori interpolation of the solution values given by the underlying discrete ODE method can be utilized. Thus, the success of the resulting DDE method in terms of accuracy and stability depends on the particular choice of the discrete method as well as of the interpolant providing the continuous extension. It can be illustrated that owing to the presence of delayed terms some desirable accuracy and stability properties of the underlying ODE method can be destroyed when the method is applied to a DDE [34]. Therefore, the integration of a DDE can rarely be based on the plain application of some classical ODE codes, rather it requires the use of specifically designed methods considering the presence of the delayed terms.

For continuous extension of the ODE method, i.e., the continuity condition of the interpolation procedure, the order of the interpolation $q \geq 1$ should be less or equal than that of the integration method [34]. In addition, it has been proven that the global order p of the piecewise discrete collocation method is preserved for any choice of the mesh by using a uniform interpolant of order $q = p - 1$ [34]. In fact, for the integration of a DDE only a few set of Runge-Kutta (RK) methods are stable. In the class of one-stage RK methods of order 1, the only one that is AN_f-stable is the backward Euler method together with linear interpolation [34]. One step collocation at one Gaussian point (the Galerkin method) is AN-stable also with linear interpolation. In the class of two-stage RK methods of order 2 the only one that is PN-stable is Lobatto IIIC method [34]. It is also shown that the DDE

$$\begin{cases} \frac{\partial}{\partial t}y(t) = \lambda(t)y(t) + \mu(t)y(t-\tau), & t_0 \leq t \\ y(t) = \phi(t), & t \leq t_0 \end{cases}$$

is PN-stable for all delay τ and initial function $\phi(t)$ in constant time step size $n\Delta t$ if $|y_n|_{n\geq 0} \leq \max|\phi(t)|$ and AN-stable if $|y_{n+1}| \leq |y_n|$.

So far, no A-stable discrete RK method of order 3 is known and it has been proven that no A-stable method can exceed order 4 [34]. In particular, no three-stage discrete RK method of order $p \geq 3$ exists which is A-stable. Hence, it is hard to find efficient high order schemes. For example, the promising schemes with averaging over the scalar potential, e.g., the Crank-Nicolson and the proposed scheme in [23], intrigued the author to check the stability of higher order integrators such as the finite difference of fourth order accuracy with three time levels, formulating (2.9) as

$$\frac{\mathbf{A}(\mathbf{r}, t_n) - \mathbf{A}(\mathbf{r}, t_{n-2})}{2\Delta t} + \frac{1}{6}[\nabla\phi(\mathbf{r}, t_n) + 4\nabla\phi(\mathbf{r}, t_{n-1}) + \nabla\phi(\mathbf{r}, t_{n-2})] = \underline{\mathbf{E}}^i(\mathbf{r}, t_{n-1}).$$

It does not, however, provide with any order of the interpolants a reliable scheme for practical purposes.

3.3 Subdomain Lagrange Basis Functions

Alternatively, to numerically solve types of the TDIEs, namely (2.10) or (2.11), and (2.24), using the Bernoulli's separation of variables, the induced surface current density $\mathbf{J}(\mathbf{r}, t)$ can be expanded in terms of the set of vector spatial and a set of scalar temporal basis functions,

$$\mathbf{J}(\mathbf{r}, t) = \sum_{n=1}^{N} \sum_{k=1}^{M} I_k^n T(t - n\Delta t) \mathbf{f}_k(\mathbf{r}). \tag{3.35}$$

Substituting (3.35) into either (2.10) or (2.11), as well as (2.24) and using Galerkin testing in space as used in Section 2.5 and applying point matching in time steps $n\Delta t$ yields to the same set of matrix equations for the unknown coefficients I_k^n as (3.16) and (3.17). Assuming that all past values of the current coefficients up to $t = t_{n-1}$ are known, those many discretized terms for which $R \geq c\Delta t$, are moved to the right-hand side of (3.16) and (3.17) and the unknown terms associated with the present time $t = t_n$ are retained on the left side. As a result, if a causal time basis function is used, i.e. $T(t) = 0$ for $t \leq -\Delta t$, (3.16) and (3.17) reduce to (3.19). Successive orders of the shifted Lagrange interpolants (each one of them ensuring the qth order approximation over the temporal subdomains) are first considered as the temporal basis function $T(t)$. In this manner, $T_n(t)$ involves $q + 1$ continuous piecewise polynomials of order q expanded over $(q + 1)\Delta t$ conjunct time intervals. These polynomials are shifted to extend over the time interval $[(n - q + 1)\Delta t, (n + 1)\Delta t]$ so that they can individually state the value of the current at time instant t depending on only one future, the present, and $q - 1$ past values of the current coefficients, as demonstrated in Fig. 3.2. In the present paper, first [1], second [5], third [8], and forth [9] -order shifted Lagrange interpolants are used as causal temporal basis functions $T(t)$ to provide respectively linear (3.36), quadratic (3.37), cubic (3.38), and quartic (3.39) approximation of the current over every generic q^{th} time interval between the samples $t_n = n\Delta t$ and $t_{n+1} = (n + 1)\Delta t$. The individual time expansion functions are defined as follows:

$$T_n(t)^{1^{\text{st}}} = \begin{cases} 1 - \frac{|t - n\Delta t|}{\Delta t}, & |t - n\Delta t| \leq \Delta t \\ 0, & \text{elsewhere} \end{cases} \tag{3.36}$$

$$T(t)^{2^{\text{nd}}} = \begin{cases} 1 \pm \frac{3}{2}(\frac{t}{\Delta t}) + \frac{1}{2}(\frac{t}{\Delta t})^2, & + : -\Delta t \leq t \leq 0,\ - : \Delta t \leq t \leq 2\Delta t \\ 1 - (\frac{t}{\Delta t})^2, & 0 \leq t \leq \Delta t \\ 0, & \text{elsewhere} \end{cases} \tag{3.37}$$

$$T(t)^{3^{\text{rd}}} = \begin{cases} 1 + \frac{11}{6}(\frac{t}{\Delta t}) + (\frac{t}{\Delta t})^2 + \frac{1}{6}(\frac{t}{\Delta t})^3, & -\Delta t \leq t \leq 0 \\ 1 + \frac{1}{2}(\frac{t}{\Delta t}) - (\frac{t}{\Delta t})^2 - \frac{1}{2}(\frac{t}{\Delta t})^3, & 0 \leq t \leq \Delta t \\ 1 - \frac{1}{2}(\frac{t}{\Delta t}) - (\frac{t}{\Delta t})^2 + \frac{1}{2}(\frac{t}{\Delta t})^3, & \Delta t \leq t \leq 2\Delta t \\ 1 - \frac{11}{6}(\frac{t}{\Delta t}) + (\frac{t}{\Delta t})^2 - \frac{1}{6}(\frac{t}{\Delta t})^3, & 2\Delta t \leq t \leq 3\Delta t \\ 0, & \text{elsewhere} \end{cases} \tag{3.38}$$

$$T(t)^{4^{\text{th}}} = \begin{cases} 1 + \frac{25}{12}(\frac{t}{\Delta t}) + \frac{35}{24}(\frac{t}{\Delta t})^2 + \frac{5}{12}(\frac{t}{\Delta t})^3 + \frac{1}{24}(\frac{t}{\Delta t})^4, & -\Delta t \leq t \leq 0 \\ 1 + \frac{5}{6}(\frac{t}{\Delta t}) - \frac{5}{6}(\frac{t}{\Delta t})^2 - \frac{5}{6}(\frac{t}{\Delta t})^3 - \frac{1}{6}(\frac{t}{\Delta t})^4, & 0 \leq t \leq \Delta t \\ 1 - \frac{5}{4}(\frac{t}{\Delta t})^2 + \frac{1}{4}(\frac{t}{\Delta t})^4, & \Delta t \leq t \leq 2\Delta t \\ 1 - \frac{5}{6}(\frac{t}{\Delta t}) - \frac{5}{6}(\frac{t}{\Delta t})^2 + \frac{5}{6}(\frac{t}{\Delta t})^3 - \frac{1}{6}(\frac{t}{\Delta t})^4, & 2\Delta t \leq t \leq 3\Delta t \\ 1 - \frac{25}{12}(\frac{t}{\Delta t}) + \frac{35}{24}(\frac{t}{\Delta t})^2 - \frac{5}{12}(\frac{t}{\Delta t})^3 + \frac{1}{24}(\frac{t}{\Delta t})^4, & 3\Delta t \leq t \leq 4\Delta t \\ 0, & \text{elsewhere} \end{cases} \tag{3.39}$$

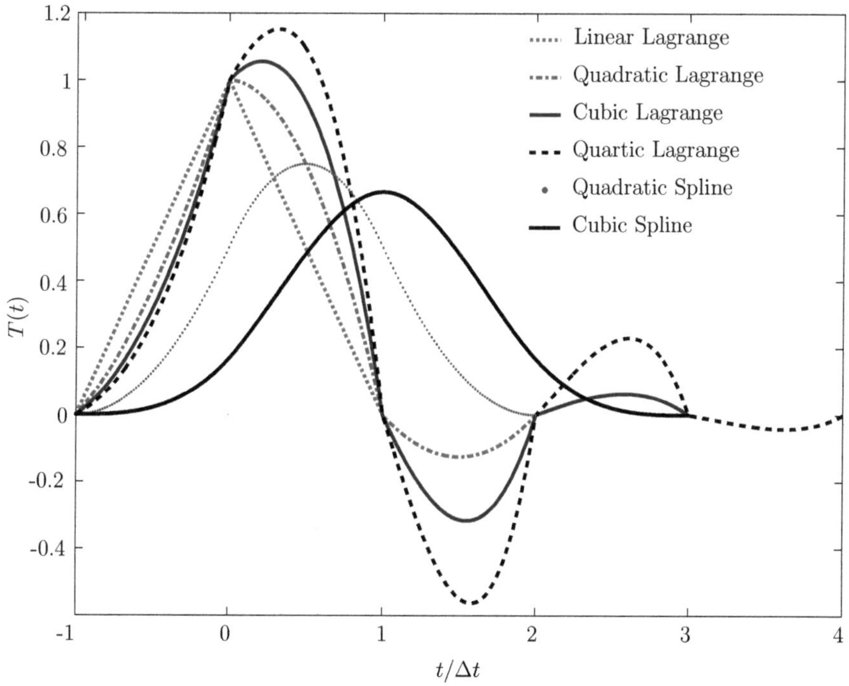

Figure 3.2: Subdomain basis functions for analytical evaluation of the time derivatives and interpolation between the retarded time samples.

In the following, as an example, the calculation of (3.1)-(3.5) for the quadratic Lagrange case (3.37) is stated:

$$\begin{aligned} a_{mk,n}^{pq} &= \sum_n I_k^n \frac{\partial T(\tau_{mk,n}^{pq} - r\Delta t)}{\partial t} = \frac{(\delta+\frac{1}{2})I_k^r - 2\delta I_k^{r-1} + (\delta-\frac{1}{2})I_k^{r-2}}{\Delta t} \\ b_{mk,n}^{pq} &= \sum_n I_k^n \int_{-\infty}^{\tau_{mk,n}^{pq}} T(t-r\Delta t)\mathrm{d}t = \Delta t\Bigg[\sum_{i=1}^{r}\left(\frac{5}{12}I_k^i + \frac{2}{3}I_k^{i-1} - \frac{1}{12}I_k^{i-2}\right) \\ &+ \left(\frac{\delta^3}{6}+\frac{\delta^2}{4}\right)I_k^r + \left(\delta - \frac{\delta^3}{3}\right)I_k^{r-1} + \left(\frac{\delta^3}{6}-\frac{\delta^2}{4}\right)I_k^{r-2}\Bigg] \end{aligned} \quad (3.40)$$

for the EFIE,

$$\begin{aligned} \dot{a}_{mk,n}^{pq} &= \sum_n I_k^n \frac{\partial^2 T(\tau_{mk,n}^{pq} - r\Delta t)}{\partial^2 t} = \frac{I_k^r - 2I_k^{r-1} + I_k^{r-2}}{\Delta t^2} \\ \dot{b}_{mk,n}^{pq} &= \sum_n I_k^n T(\tau_{mk,n}^{pq} - r\Delta t) \\ &= \frac{\delta(\delta+1)}{2}I_k^r + (1-\delta^2)I_k^{r-1} + \frac{\delta(\delta-1)}{2}I_k^{r-2} \end{aligned} \quad (3.41)$$

for the DEFIE, and

$$\begin{aligned} c_{mk,n}^{pq} &= \sum_n I_k^n \frac{\partial T(\tau_{mk,n}^{pq} - r\Delta t)}{\partial t} = a_{mk,n}^{pq} \\ d_{mk,n}^{pq} &= \sum_n I_k^n T(\tau_{mk,n}^{pq} - r\Delta t) = \dot{b}_{mk,n}^{pq} \\ f_{mk,n}^{pq} &= \sum_n I_k^n T(t_n - n\Delta t) = I_k^n \end{aligned}$$

for the MFIE. Similarly for the other choices of $T(t)$, the closed-form expression of the selected time bases $T(t)$ is directly utilized for analytical evaluation of the time derivatives in the TDIE and for interpolation of the current values at the retarded times. It is worth mentioning that other sets of basis functions with a simple closed-form expression, short temporal support, total positivity, and the desired unit sum property $\sum_{n=-\infty}^{+\infty} T(t-n\Delta t) = 1$ were implemented. In this regard, the cosine squared function Fig. 3.3(a) [18],

$$T_n(t) = \begin{cases} \cos^2(\frac{\pi}{2}\frac{t-n\Delta t}{\Delta t}), & |t-n\Delta t| \leq \Delta t \\ 0, & \text{elsewhere} \end{cases},$$

the optimized exponential function with all-order continuous derivatives, Fig. 3.4(a), [19],

$$T(t) = \begin{cases} \exp[-\frac{4.6487(\frac{t}{\Delta t})^2}{(1-(\frac{t}{\Delta t})^2)(1+5(\frac{t}{\Delta t})^2)}], & |t| \leq \Delta t \\ 0, & \text{elsewhere} \end{cases},$$

a causal truncated form (Fig. 3.3(b)) of the approximate prolate spheroidal wave [6][1],

$$T(t) = \frac{\sin(s\omega_0 t)}{s\omega_0 t}\frac{\sin[a\sqrt{(\frac{t}{N\Delta t})^2-1}]}{\sinh(a)\sqrt{(\frac{t}{N\Delta t})^2-1}} \qquad a = \pi N\frac{s-1}{s}, \text{integer} \quad N \geq 1 \quad (3.42)$$

[1] a is called the time-bandwidth product for the oversampling factor $10 \leq s \leq 20$, and ω_0 is the highest frequency in the band of interest.

and the cubic Hermite spline functions, Fig. 3.3(a),

$$T(t) = \begin{cases} (1+\frac{t}{\Delta t})^2(1-2\frac{t}{\Delta t}), & -\Delta t \leq t \leq 0 \\ (1-\frac{t}{\Delta t})^2(1+2\frac{t}{\Delta t}), & 0 \leq t \leq \Delta t \\ 0, & \text{elsewhere} \end{cases},$$

were also tested to analytically evaluate the time derivatives and interpolate the delay terms. It was, however, observed that the incorporation of any of them causes numerical instability at very early time stages of the (D)EFIE solution. The optimized sinusoidal dome interpolant Fig. 3.4(b) proposed by [23]

$$T(t) = \begin{cases} [\frac{\sin(\pi|\frac{t}{\Delta t}|^{0.3})}{\pi|\frac{t}{\Delta t}|^{0.3}}]^{0.462}, & |t| \leq \Delta t \\ 0, & \text{elsewhere} \end{cases},$$

is not differentiable at the integer multiples of Δt. The noncausal approximate prolate spheroidal wave (Knab's) functions (3.42) violate the marching condition since the current value at a given time step relies on its value in future time steps [6, 66], and extrapolation of present and past currents may not be able to predict future currents accurately, especially at large time steps. Note that [24] recently has reported successful use of quadratic B-spline basis functions as explained above but stabilized in the CFIE framework.

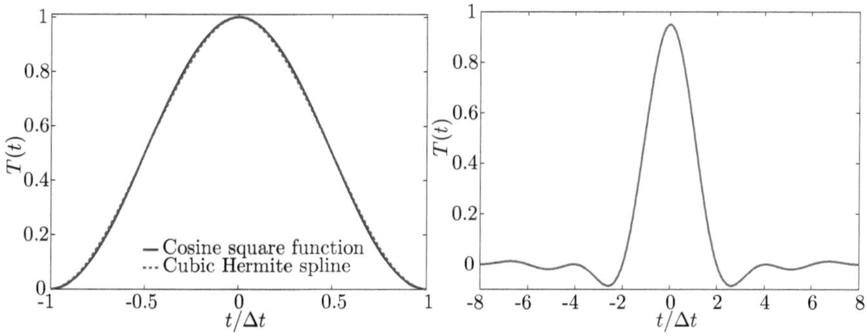

(a) Cosine squared and cubic Hermite spline functions

(b) Truncated bandlimited approximate prolate spheroidal wave functions of Knab

Figure 3.3: Unsuitable choices of time basis functions (a) instable when derivatives are handled analytically; (b) needs extrapolation.

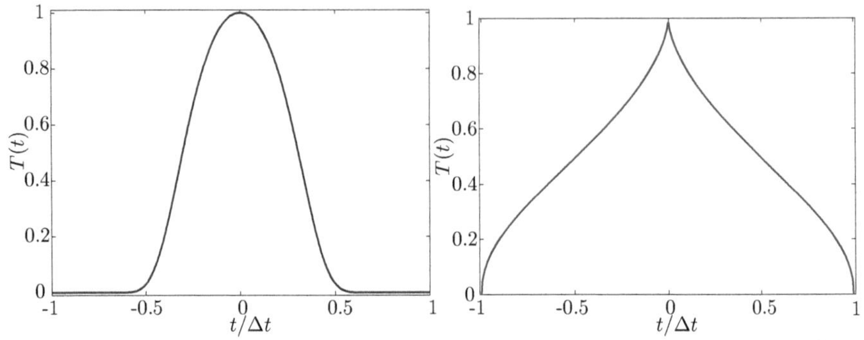

(a) Optimized exponential function with all-order continuous derivatives (b) Optimized sinusoidal dome interpolation function

Figure 3.4: Artificial time interpolation functions.

3.3.1 B-Spline Bases with Entire-Domain Interpolation

As alternative choices, two new temporal basis functions are proposed [36], the quadratic and cubic cardinal B-spline function, respectively, defined as

$$T(t) = \begin{cases} \frac{1}{2} + (\frac{t}{\Delta t}) + \frac{1}{2}(\frac{t}{\Delta t})^2, & -\Delta t \leq t \leq 0 \\ \frac{1}{2} + (\frac{t}{\Delta t}) - (\frac{t}{\Delta t})^2, & 0 \leq t \leq \Delta t \\ 2 - 2(\frac{t}{\Delta t}) + \frac{1}{2}(\frac{t}{\Delta t})^2, & \Delta t \leq t \leq 2\Delta t \\ 0, & \text{elsewhere} \end{cases} \tag{3.43}$$

$$T(t) = \begin{cases} -1 + 3(\frac{t}{\Delta t}) - 3(\frac{t}{\Delta t})^2 + (\frac{t}{\Delta t})^3, & -\Delta t \leq t \leq 0 \\ 4 - 12(\frac{t}{\Delta t}) + 12(\frac{t}{\Delta t})^2 - 3(\frac{t}{\Delta t})^3, & 0 \leq t \leq \Delta t \\ -5 + 21(\frac{t}{\Delta t}) - 15(\frac{t}{\Delta t})^2 + 3(\frac{t}{\Delta t})^3, & \Delta t \leq t \leq 2\Delta t \\ 8 - 12(\frac{t}{\Delta t}) + 6(\frac{t}{\Delta t})^2 - (\frac{t}{\Delta t})^3, & 2\Delta t \leq t \leq 3\Delta t \\ 0, & \text{elsewhere.} \end{cases} \tag{3.44}$$

Analytical evaluation of the time derivatives in the TDIEs can be derived directly through (3.43) or (3.44) and for interpolation of the current values at the retarded times the cubic spline interpolation is exploited [72]. The cubic spline interpolation formula provides an interpolant that is smooth in the first derivative and continuous in the second derivative within all time intervals and their boundaries. For a unique solution, the spline interpolation routine needs two further (boundary) conditions. Here, the natural cubic spline interpolation is used that specifies zero second derivatives on both the initial and the present time instances, $\left.\frac{\partial^2 \mathbf{J}(\mathbf{r},t)}{\partial t^2}\right|_{t=t_0,t_n} = 0$. For the sake of computational efficiency of the entire domain interpolation routine, the required second derivatives of the function at sample points are estimated once and saved for further usages [72].

3.4 Entire-Domain Basis Functions

To numerically solve the TDIE (2.10), (2.11), or (2.24) in a marching recipe, another choice is using the spatial vector basis functions in conjunction with a set of entire-domain but causal temporal coefficient functions $c_k(t)$, whereby the induced surface current is approximately expanded by

$$\mathbf{J}(\mathbf{r},t) = \sum_{k=1}^{M} c_k(t)\mathbf{f}_k(\mathbf{r}) \tag{3.45}$$

and the transient coefficients

$$c_k(t) = \sum_{j=0}^{\infty} c_{k,j}\phi_j(st) \tag{3.46}$$

are weighted Laguerre polynomials $\phi_j(st) = e^{-st/2}L_j(st)$. The Laguerre polynomial of order j, $L_j(st)$, can be recursively regenerated in a stable manner from their lower orders

$$\begin{aligned} L_0(t) &= 1, \; L_1(t) = 1 - t \\ jL_j(t) &= (2j-1-t)L_{j-1}(t) - (j-1)L_{j-2}(t) \qquad \text{for } j \geq 2,\, t \geq 0. \end{aligned}$$

An advantageous feature of the Laguerre polynomials is that they are orthogonal with respect to the kernel e^{-t}, that is

$$\int_0^{\infty} e^{-t}L_i(t)L_j(t)\mathrm{d}t = \delta_{ij} \tag{3.47}$$

or equivalently for the weighted Laguerre polynomials

$$\int_0^{\infty} \phi_i(st)\phi_j(st)\mathrm{d}(st) = \delta_{ij}$$

where δ_{ij} is the Kronecker delta, that is one for $i = j$ and zero otherwise. Therefore, $\phi_j(st)$ $j = 0, 1, 2, \ldots$ provide a complete orthogonal set of basis functions in $L^2(R^+)$ that decay to zero as $t \to \infty$, as exhibited in Fig. 3.5. The scaling factor s controls the temporal support provided by the expansion. The larger the scaling factor is, the finer the resolution in time is. The closed-form analytical expressions for the first and the second derivatives as well as the indefinite integral of the transient coefficients (3.46) with respect to time are available (see Appendix 8.4)

$$\begin{aligned} \frac{\mathrm{d}}{\mathrm{d}t}c_k(t) &= s\sum_{j=0}^{\infty}\left(\frac{1}{2}c_{k,j} + \sum_{\iota=0}^{j-1}c_{k,\iota}\right)\phi_j(st) && (3.48) \\ \frac{\mathrm{d}^2}{\mathrm{d}t^2}c_k(t) &= s^2\sum_{j=0}^{\infty}\left[\frac{1}{4}c_{k,j} + \sum_{\iota=0}^{j-1}(j-\iota)c_{k,\iota}\right]\phi_j(st) && (3.49) \\ \int_0^t c_k(\tau)\mathrm{d}\tau &= \frac{2}{s}\sum_{j=0}^{\infty}\left[c_{k,j} + 2\sum_{\iota=0}^{j-1}(-1)^{j+\iota}c_{k,\iota}\right]\phi_j(st). && (3.50) \end{aligned}$$

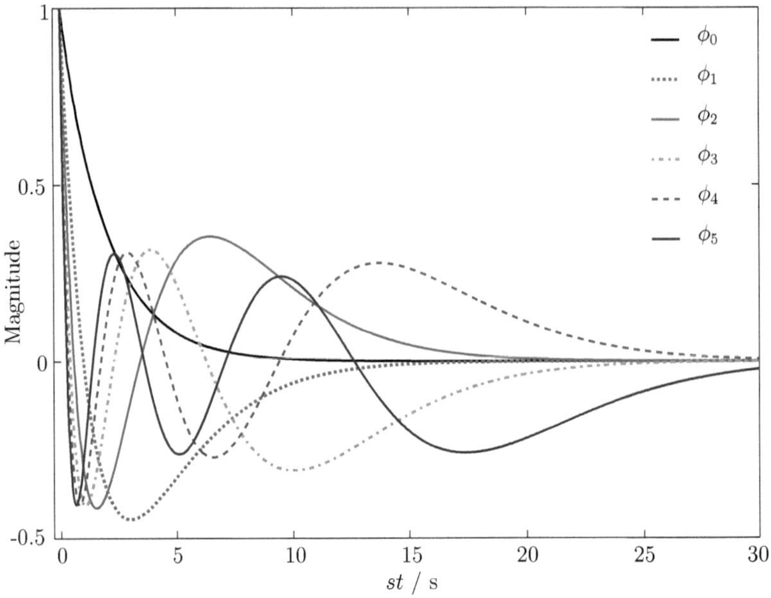

Figure 3.5: Weighted Laguerre polynomials of different orders.

3.4.1 Laguerre Expansion Method

Substituting (3.46) into (3.45)

$$\mathbf{J}(\mathbf{r},t) = \sum_{k=1}^{M}\sum_{j=0}^{\infty} c_{k,j}\phi_j(st)\mathbf{f}_k(\mathbf{r}) \tag{3.51}$$

and subsequently inserting it in the EFIE (2.10) using (3.48) and (3.50), one obtains

$$\frac{\mu s}{4\pi}\sum_{k=1}^{M}\sum_{j=0}^{\infty}\left(\frac{1}{2}c_{k,j} + \sum_{\iota=0}^{j-1} c_{k,\iota}\right)\int_S \frac{\phi_j(s\tau)\mathbf{f}_k(\mathbf{r}')}{R}\mathrm{d}S'$$
$$-\frac{1}{2\pi\epsilon s}\sum_{k=1}^{M}\sum_{j=0}^{\infty}\left[c_{k,j} + 2\sum_{\iota=0}^{j-1}(-1)^{j+\iota}c_{k,\iota}\right]\int_S \nabla_{\mathbf{r}}\left[\frac{\phi_j(s\tau)\nabla_{\mathbf{r}'}\cdot\mathbf{f}_k(\mathbf{r}')}{R}\right]\mathrm{d}S' = \underline{\mathbf{E}}^i(\mathbf{r},t). \tag{3.52}$$

Likewise, expanding the unknowns of the DEFIE (2.11) by (3.45) and using (3.49) results in

$$\frac{\mu s^2}{4\pi}\sum_{k=1}^{M}\sum_{j=0}^{\infty}\left[\frac{1}{4}c_{k,j} + \sum_{\iota=0}^{j-1}(j-\iota)c_{k,\iota}\right]\int_S \frac{\phi_j(s\tau)\mathbf{f}_k(\mathbf{r}')}{R}\mathrm{d}S'$$
$$-\frac{1}{4\pi\epsilon}\sum_{k=1}^{M}\sum_{j=0}^{\infty} c_{k,j}\int_S \nabla_{\mathbf{r}}\left[\frac{\phi_j(s\tau)\nabla_{\mathbf{r}'}\cdot\mathbf{f}_k(\mathbf{r}')}{R}\right]\mathrm{d}S' = \frac{\partial\underline{\mathbf{E}}^i(\mathbf{r},t)}{\partial t}, \tag{3.53}$$

and as the third and last case, substituting (3.46) again into (3.45) and inserting it in the MFIE (2.24) using (3.48), we have

$$\sum_{k=1}^{M}\sum_{j=0}^{\infty}\frac{1}{2}c_{k,j}\phi_j(st)\mathbf{f}_k(\mathbf{r}')$$
$$-\frac{1}{4\pi}\hat{\mathbf{n}}\times\left[\frac{s}{c}\sum_{k=1}^{M}\sum_{j=0}^{\infty}\left(\frac{1}{2}c_{k,j}+\sum_{\iota=0}^{j-1}c_{k,\iota}\right)\int_{S_0}\frac{\phi_j(s\tau)\mathbf{f}_k(\mathbf{r}')\times\mathbf{R}}{R^2}\mathrm{d}S'\right.$$
$$\left.+\sum_{k=1}^{M}\sum_{j=0}^{\infty}c_{k,j}\int_{S_0}\frac{\phi_j(s\tau)\mathbf{f}_k(\mathbf{r}')\times\mathbf{R}}{R^3}\mathrm{d}S'\right]=\hat{\mathbf{n}}\times\mathbf{H}^i(\mathbf{r},t). \tag{3.54}$$

Performing the Galerkin's testing procedure for the spatial expansion functions by the same vector valued $\mathbf{f}_m(\mathbf{r})$, $m=1,2,\ldots,M$, (3.52), (3.53), and (3.54), respectively, give

$$\frac{\mu s}{4\pi}\sum_{k=1}^{M}\sum_{j=0}^{\infty}\left(\frac{1}{2}c_{k,j}+\sum_{\iota=0}^{j-1}c_{k,\iota}\right)\int_S\mathbf{f}_m(\mathbf{r})\cdot\int_S\frac{\phi_j(s\tau)\mathbf{f}_k(\mathbf{r}')}{R}\mathrm{d}S'\mathrm{d}S$$
$$+\frac{1}{2\pi\epsilon s}\sum_{k=1}^{M}\sum_{j=0}^{\infty}\left[c_{k,j}+2\sum_{\iota=0}^{j-1}(-1)^{j+\iota}c_{k,\iota}\right]$$
$$\int_S\nabla_{\mathbf{r}}\cdot\mathbf{f}_m(\mathbf{r})\int_S\left[\frac{\phi_j(s\tau)\nabla_{\mathbf{r}'}\cdot\mathbf{f}_k(\mathbf{r}')}{R}\right]\mathrm{d}S'\mathrm{d}S=\int_S\mathbf{f}_m(\mathbf{r})\cdot\underline{\mathbf{E}}^i(\mathbf{r},t)\mathrm{d}S \tag{3.55}$$

$$\frac{\mu s^2}{4\pi}\sum_{k=1}^{M}\sum_{j=0}^{\infty}\left[\frac{1}{4}c_{k,j}+\sum_{\iota=0}^{j-1}(j-\iota)c_{k,\iota}\right]\int_S\mathbf{f}_m(\mathbf{r})\cdot\int_S\frac{\phi_j(s\tau)\mathbf{f}_k(\mathbf{r}')}{R}\mathrm{d}S'\mathrm{d}S$$
$$+\frac{1}{4\pi\epsilon}\sum_{k=1}^{M}\sum_{j=0}^{\infty}c_{k,j}\int_S\nabla_{\mathbf{r}}\cdot\mathbf{f}_m(\mathbf{r})\int_S\left[\frac{\phi_j(s\tau)\nabla_{\mathbf{r}'}\cdot\mathbf{f}_k(\mathbf{r}')}{R}\right]\mathrm{d}S'\mathrm{d}S=\int_S\mathbf{f}_m(\mathbf{r})\cdot\frac{\partial\underline{\mathbf{E}}^i(\mathbf{r},t)}{\partial t}\mathrm{d}S \tag{3.56}$$

$$\sum_{k=1}^{M}\sum_{j=0}^{\infty}\frac{1}{2}c_{k,j}\phi_j(st)\int_S\mathbf{f}_m(\mathbf{r})\cdot\mathbf{f}_k(\mathbf{r}')\mathrm{d}S$$
$$-\left[\frac{s}{4\pi c}\sum_{k=1}^{M}\sum_{j=0}^{\infty}\left(\frac{1}{2}c_{k,j}+\sum_{\iota=0}^{j-1}c_{k,\iota}\right)\int_S\mathbf{f}_m(\mathbf{r})\cdot\hat{\mathbf{n}}\times\int_S\frac{\phi_j(s\tau)\mathbf{f}_k(\mathbf{r}')\times\mathbf{R}}{R^2}\mathrm{d}S'\mathrm{d}S\right.$$
$$\left.+\frac{1}{4\pi}\sum_{k=1}^{M}\sum_{j=0}^{\infty}c_{k,j}\int_S\mathbf{f}_m(\mathbf{r})\cdot\hat{\mathbf{n}}\times\int_S\frac{\phi_j(s\tau)\mathbf{f}_k(\mathbf{r}')\times\mathbf{R}}{R^3}\mathrm{d}S'\mathrm{d}S\right]$$
$$=\int_S\mathbf{f}_m(\mathbf{r})\cdot\hat{\mathbf{n}}\times\mathbf{H}^i(\mathbf{r},t)\mathrm{d}S. \tag{3.57}$$

Finally, following the spatial testing, the temporal testing is applied in Galerkin context respectively to (3.55), (3.56), and (3.57) using $\phi_i(st)$ ($i=0,1,2,\ldots,N$) as the weighting functions. This results in recursive relations between the different orders of the Laguerre polynomials

$$\frac{\mu s}{4\pi}\sum_{k=1}^{M}\sum_{j=0}^{i}\left(\frac{1}{2}c_{k,j}+\sum_{\iota=0}^{j-1}c_{k,\iota}\right)\int_S \mathbf{f}_m(\mathbf{r})\cdot\int_S I_\nu(s\frac{R}{c})\frac{\mathbf{f}_k(\mathbf{r}')}{R}\mathrm{d}S'\mathrm{d}S+\frac{1}{2\pi\epsilon s}$$
$$\sum_{k=1}^{M}\sum_{j=0}^{i}\left[c_{k,j}+2\sum_{\iota=0}^{j-1}(-1)^{j+\iota}c_{k,\iota}\right]\int_S \nabla_{\mathbf{r}}\cdot\mathbf{f}_m(\mathbf{r})\int_S\left[I_\nu(s\frac{R}{c})\frac{\nabla_{\mathbf{r}'}\cdot\mathbf{f}_k(\mathbf{r}')}{R}\right]\mathrm{d}S'\mathrm{d}S=e_{m,i}\tag{3.58}$$

$$\frac{\mu s^2}{4\pi}\sum_{k=1}^{M}\sum_{j=0}^{i}\left[\frac{1}{4}c_{k,j}+\sum_{\iota=0}^{j-1}(j-\iota)c_{k,\iota}\right]\int_S \mathbf{f}_m(\mathbf{r})\cdot\int_S I_\nu(s\frac{R}{c})\frac{\mathbf{f}_k(\mathbf{r}')}{R}\mathrm{d}S'\mathrm{d}S$$
$$+\frac{1}{4\pi\epsilon}\sum_{k=1}^{M}\sum_{j=0}^{i}c_{k,j}\int_S \nabla_{\mathbf{r}}\cdot\mathbf{f}_m(\mathbf{r})\int_S\left[I_\nu(s\frac{R}{c})\frac{\nabla_{\mathbf{r}'}\cdot\mathbf{f}_k(\mathbf{r}')}{R}\right]\mathrm{d}S'\mathrm{d}S=\dot{e}_{m,i}\tag{3.59}$$

$$\sum_{k=1}^{M}\sum_{j=0}^{i}\frac{1}{2}c_{k,i}\delta_{ij}\int_S \mathbf{f}_m(\mathbf{r})\cdot\mathbf{f}_k(\mathbf{r}')\mathrm{d}S$$
$$-\left[\frac{s}{4\pi c}\sum_{k=1}^{M}\sum_{j=0}^{i}\left(\frac{1}{2}c_{k,j}+\sum_{\iota=0}^{j-1}c_{k,\iota}\right)\int_S \mathbf{f}_m(\mathbf{r})\cdot\hat{\mathbf{n}}\times\int_S I_\nu(s\frac{R}{c})\frac{\mathbf{f}_k(\mathbf{r}')\times\mathbf{R}}{R^2}\mathrm{d}S'\mathrm{d}S\right.$$
$$\left.+\frac{1}{4\pi}\sum_{k=1}^{M}\sum_{j=0}^{i}c_{k,j}\int_S \mathbf{f}_m(\mathbf{r})\cdot\hat{\mathbf{n}}\times\int_S I_\nu(s\frac{R}{c})\frac{\mathbf{f}_k(\mathbf{r}')\times\mathbf{R}}{R^3}\mathrm{d}S'\mathrm{d}S\right]=h_{m,i},\tag{3.60}$$

where

$$e_{m,i} = \int_0^\infty \phi_i(st)\int_S \mathbf{f}_m(\mathbf{r})\cdot\underline{\mathbf{E}}^i(\mathbf{r},t)\mathrm{d}S\,\mathrm{d}(st)\tag{3.61}$$
$$\dot{e}_{m,i} = \int_0^\infty \phi_i(st)\int_S \mathbf{f}_m(\mathbf{r})\cdot\frac{\partial\underline{\mathbf{E}}^i(\mathbf{r},t)}{\partial t}\,\mathrm{d}S\,\mathrm{d}(st)\tag{3.62}$$
$$h_{m,i} = \int_0^\infty \phi_i(st)\int_S \mathbf{f}_m(\mathbf{r})\cdot\hat{\mathbf{n}}\times\mathbf{H}^i(\mathbf{r},t)\mathrm{d}S\,\mathrm{d}(st),\tag{3.63}$$

and $\nu = i - j$. Considering the orthogonality of the temporal basis function (3.47) and using the Sheffer's identity in 8.4.1, we define

$$I_\nu(s\frac{R}{c})=\int_0^\infty \phi_i(st)\phi_j(st-s\frac{R}{c})\mathrm{d}(st)=\begin{cases} e^{-\frac{sR}{2c}}\left[L_\nu(s\frac{R}{c})-L_{\nu-1}(s\frac{R}{c})\right] & \nu>0\\ e^{-\frac{sR}{2c}} & \nu=0\\ 0 & \nu<0\end{cases}\tag{3.64}$$

or concisely

$$I_\nu(s\frac{R}{c})=\phi_\nu(s\frac{R}{c})-\phi_{\nu-1}(s\frac{R}{c})$$

where $\phi_\nu = 0$ for $\nu < 0$. This allows to simplify the left-hand side of (3.58)-(3.60) and modify the upper bound of the order summation index from ∞ in (3.55)-(3.57) to i.

3.4.2 Marching-on-in-Degree (MOD) Recipes

To handle the retarded terms, again we assume the unknown transient quantity does not appreciably change within the triangle,

$$I_\nu(s\frac{R}{c}) \longrightarrow I_\nu(s\frac{R^{pq}_{mk}}{c}). \tag{3.65}$$

Thus, all the previously derived relations (3.1)-(3.7) can be also used here provided that n is interchanged with ν for (3.58), (3.59), and (3.60). Therefore, one obtains

$$\begin{aligned} a^{pq}_{mk,\nu} &= sI_\nu(s\frac{R^{pq}_{mk}}{c}) \\ b^{pq}_{mk,\nu} &= \frac{2}{s}I_\nu(s\frac{R^{pq}_{mk}}{c}) \end{aligned}$$

$$\sum_{k=1}^{M}\left[\sum_{j=0}^{i}\left(\frac{a_{mk,\nu}}{2}+b_{mk,\nu}\right)c_{k,j}+\sum_{j=0}^{i}\sum_{\iota=0}^{j-1}\left(a_{mk,\nu}+2(-1)^{j+\iota}b_{mk,\nu}\right)c_{k,\iota}\right]=e_{m,i} \tag{3.66}$$

for the EFIE case (3.58), while using

$$\begin{aligned} \dot{a}^{pq}_{mk,\nu} &= s^2I_\nu(s\frac{R^{pq}_{mk}}{c}) \\ \dot{b}^{pq}_{mk,\nu} &= I_\nu(s\frac{R^{pq}_{mk}}{c}) \end{aligned}$$

results in

$$\sum_{k=1}^{M}\left[\sum_{j=0}^{i}\left(\frac{\dot{a}_{mk,\nu}}{4}+\dot{b}_{mk,\nu}\right)c_{k,j}+\sum_{j=0}^{i}\sum_{\iota=0}^{j-1}(j-\iota)\dot{a}_{mk,\nu}c_{k,\iota}\right]=\dot{e}_{m,i} \tag{3.67}$$

for the DEFIE (3.59), and specifying

$$\begin{aligned} c^{pq}_{mk,\nu} &= sI_\nu(s\frac{R^{pq}_{mk}}{c}) \\ d^{pq}_{mk,\nu} &= I_\nu(s\frac{R^{pq}_{mk}}{c}) \\ f^{pq}_{mk,i} &= 1 \end{aligned}$$

the discretized MFIE (3.60) can be represented by

$$\sum_{k=1}^{M}\left[f_{mk,i}c_{k,i}-\sum_{j=0}^{i}\left(\frac{c_{mk,\nu}}{2}+d_{mk,\nu}\right)c_{k,j}-\sum_{j=0}^{i}\sum_{\iota=0}^{j-1}c_{mk,\nu}\,c_{k,\iota}\right]=h_{m,i}. \tag{3.68}$$

Assuming that all the lower orders of the expansion coefficients up to $i-1$ are known, they are moved to the right-hand sides of (3.66), (3.67), and (3.68) whereas the coefficients associated with the present order $j = i$ are retained on the left-hand sides to shape the final system of equations:

$$\sum_{k=1}^{M}\left(\frac{a_{mk,0}}{2}+b_{mk,0}\right)c_{k,i} = e_{m,i} - \sum_{j=0}^{i-1}\sum_{k=1}^{M}\left(\frac{a_{mk,\nu}}{2}+b_{mk,\nu}\right)c_{k,j}$$
$$-\sum_{j=0}^{i}\sum_{\iota=0}^{j-1}\sum_{k=1}^{M}\left(a_{mk,\nu}+2(-1)^{j+\iota}b_{mk,\nu}\right)c_{k,\iota} \qquad m=1,2,\ldots,M \tag{3.69}$$

$$\sum_{k=1}^{M}\left(\frac{\dot{a}_{mk,0}}{4}+\dot{b}_{mk,0}\right)c_{k,i} = \dot{e}_{m,i} - \sum_{j=0}^{i-1}\sum_{k=1}^{M}\left(\frac{\dot{a}_{mk,\nu}}{4}+\dot{b}_{mk,\nu}\right)c_{k,j}$$
$$-\sum_{j=0}^{i}\sum_{\iota=0}^{j-1}\sum_{k=1}^{M}(j-\iota)\dot{a}_{mk,\nu}c_{k,\iota} \qquad m=1,2,\ldots,M \tag{3.70}$$

$$\sum_{k=1}^{M}\left[f_{mk,0}-\left(\frac{c_{mk,0}}{2}+d_{mk,0}\right)\right]c_{k,i} = h_{m,i} + \sum_{j=0}^{i-1}\sum_{k=1}^{M}\left(\frac{c_{mk,\nu}}{2}+d_{mk,\nu}\right)c_{k,j}$$
$$+\sum_{j=0}^{i}\sum_{\iota=0}^{j-1}\sum_{k=1}^{M}c_{mk,\nu}\,c_{k,\iota} \qquad m=1,2,\ldots,M. \tag{3.71}$$

In the above MOD formulations (3.69)-(3.71), regardless of all previously introduced MOD types, the summations over the contributions of all spatial sources $\sum_k$ are applied before the accumulations of the expansion orders $\sum_j$. This space-order summation interchange facilitates later on the implementation of acceleration schemes described in Chapter 4. Finally, one can summarize the solution procedure for the EFIE (3.69) or the DEFIE (3.70) and the MFIE (3.71) into matrix-equation forms as

$$\bar{\bar{\mathrm{Z}}}_0\bar{\mathrm{I}}_i = \bar{\mathrm{V}}_i - \sum_{j=0}^{i-1}\bar{\bar{\mathrm{Z}}}_\nu\bar{\mathrm{I}}_j - \sum_{j=0}^{i}\sum_{\iota=0}^{j-1}(\bar{\bar{\mathrm{Z}}}_\nu^{\mathrm{A}}+(-1)^{j+\iota}\bar{\bar{\mathrm{Z}}}_\nu^{\phi})\bar{\mathrm{I}}_\iota \qquad i=0,1,2,\ldots,N \tag{3.72}$$

$$\dot{\bar{\bar{\mathrm{Z}}}}_0\bar{\mathrm{I}}_i = \dot{\bar{\mathrm{V}}}_i - \sum_{j=0}^{i-1}\dot{\bar{\bar{\mathrm{Z}}}}_\nu\bar{\mathrm{I}}_j - \sum_{j=0}^{i}\sum_{\iota=0}^{j-1}(j-\iota)\bar{\bar{\mathrm{Z}}}_\nu^{\Phi}\bar{\mathrm{I}}_\iota \qquad i=0,1,2,\ldots,N \tag{3.73}$$

$$\bar{\bar{\mathrm{Z}}}_0\bar{\mathrm{I}}_i = \bar{\mathrm{V}}_i + \sum_{j=0}^{i-1}\bar{\bar{\mathrm{Z}}}_\nu\bar{\mathrm{I}}_j + \sum_{j=0}^{i}\sum_{\iota=0}^{j-1}\bar{\bar{\mathrm{Z}}}_\nu^{A}\bar{\mathrm{I}}_\iota \qquad i=0,1,2,\ldots,N \tag{3.74}$$

Since the coefficient matrix $\bar{\bar{\mathrm{Z}}}_0$ is independent of the order of the temporal testing function, the resulting dense linear matrix equation can be solved recursively by performing the LU decomposition once at the first iteration and then using the back-substitution technique repeatedly on demand. To solve the CFIE by the MOD scheme, (3.72) is augmented by $\eta\frac{(1-\kappa)}{\kappa}$ fraction of (3.74) as elucidated by (2.25). The minimum number of the temporal basis functions N can be determined by the time-bandwidth product of the response. Readers are referred to the appendix 8.5.1 for the details.

3.4.3 Advanced Marching-on-in-Degree (AMOD) Methods

To numerically solve (2.10), (2.11), (2.24) or any linear combinations of them, after inserting the unknown expansion (3.45) into the integral equations, the next stages of discretization involve testing the TDIE by the same set of spatial vector and scalar temporal basis functions. Any either of the spatial and temporal testing can be accomplished before the other. In contrary to Section 3.4.1, the spatial testing is applied here followed by the temporal testing [43], however, ultimately the accumulations of the expansion orders $\sum_j$ are rearranged exterior to the summations over the contributions of all spatial source subdomains $\sum_k$ as in Section 4.2.

Inserting (3.46) in (3.45) and then substituting the result into respectively (2.10), (2.11), (2.24) by applying the temporal testing in Galerkin sense, i.e., taking the inner product of them with the same weighting functions as the expansion ones $< \phi_i(st), \cdot >$, the time variable is integrated out:

$$\frac{\mu s}{4\pi}\sum_{k=1}^{M}\sum_{j=0}^{i}\left[\frac{1}{2}c_{k,j}+\sum_{\iota=0}^{j-1}c_{k,\iota}\right]\int_S I_\nu(s\frac{R}{c})\frac{\mathbf{f}_k(\mathbf{r}')}{R}\mathrm{d}S' - \frac{2}{4\pi\epsilon s}\sum_{k=1}^{M}\sum_{j=0}^{i}\left[c_{k,j}+2\sum_{\iota=0}^{j-1}(-1)^{j+\iota}c_{k,\iota}\right]$$
$$\int_S \nabla_{\mathbf{r}}\left[I_\nu(s\frac{R}{c})\frac{\nabla_{\mathbf{r}'}\cdot\mathbf{f}_k(\mathbf{r}')}{R}\right]\mathrm{d}S' = \int_0^\infty \phi_i(st)\underline{\mathbf{E}}^i(\mathbf{r},t)\mathrm{d}(st) \quad (3.75)$$

$$\frac{\mu s^2}{4\pi}\sum_{k=1}^{M}\sum_{j=0}^{i}\left[\frac{1}{4}c_{k,j}+\sum_{\iota=0}^{j-1}(j-\iota)c_{k,\iota}\right]\int_S I_\nu(s\frac{R}{c})\frac{\mathbf{f}_k(\mathbf{r}')}{R}\mathrm{d}S'$$
$$-\frac{1}{4\pi\epsilon}\sum_{k=1}^{M}\sum_{j=0}^{i}c_{k,j}\int_S \nabla_{\mathbf{r}}\left[I_\nu(s\frac{R}{c})\frac{\nabla_{\mathbf{r}'}\cdot\mathbf{f}_k(\mathbf{r}')}{R}\right]\mathrm{d}S' = \int_0^\infty \phi_i(st)\frac{\partial\underline{\mathbf{E}}^i(\mathbf{r},t)}{\partial t}\mathrm{d}(st) \quad (3.76)$$

$$\sum_{k=1}^{M}\sum_{j=0}^{i}\frac{1}{2}c_{k,j}\delta_{ij}\mathbf{f}_k(\mathbf{r}') - \hat{\mathbf{n}}\times\frac{1}{4\pi}\left[\frac{s}{c}\sum_{k=1}^{M}\sum_{j=0}^{i}\left(\frac{1}{2}c_{k,j}+\sum_{\iota=0}^{j-1}c_{k,\iota}\right)\int_{S_0} I_\nu(s\frac{R}{c})\frac{\mathbf{f}_k(\mathbf{r}')\times\mathbf{R}}{R^2}\mathrm{d}S'\right.$$
$$\left.+\sum_{k=1}^{M}\sum_{j=0}^{i}c_{k,j}\int_{S_0} I_\nu(s\frac{R}{c})\frac{\mathbf{f}_k(\mathbf{r}')\times\mathbf{R}}{R^3}\mathrm{d}S'\right] = \int_0^\infty \phi_i(st)\hat{\mathbf{n}}\times\mathbf{H}^i(\mathbf{r},t)\mathrm{d}(st) \quad (3.77)$$

where the orthogonality of the temporal basis function (3.47) facilitates the definition of I_{i-j} (3.64) and eventually permits to modify the upper limit of the sums over the polynomial orders from ∞ in (3.46) to i in (3.75)-(3.77). Following the temporal testing, by applying the spatial testing $< \mathbf{f}_m, \cdot >$ on (3.75)-(3.77) for all $m = 1, 2, \ldots, M$, the obtained recursive relations between different orders of the Laguerre polynomials are formulated respectively as the matrix equations (3.69)-(3.71) with the following coefficient matrices and excitation vectors:

$$a_{mk,\nu} = \frac{\mu s}{4\pi}\int_S \mathbf{f}_m(\mathbf{r})\cdot\int_S I_\nu(s\frac{R}{c})\frac{\mathbf{f}_k(\mathbf{r}')}{R}\mathrm{d}S'\mathrm{d}S \quad (3.78)$$

$$b_{mk,\nu} = \frac{1}{2\pi\epsilon s}\int_S \nabla_{\mathbf{r}}\cdot\mathbf{f}_m(\mathbf{r})\int_S I_\nu(s\frac{R}{c})\frac{\nabla_{\mathbf{r}'}\cdot\mathbf{f}_k(\mathbf{r}')}{R}\mathrm{d}S'\mathrm{d}S \quad (3.79)$$

$$c_{mk,\nu} = \frac{s}{4\pi c}\int_S \mathbf{f}_m(\mathbf{r}) \cdot \hat{\mathbf{n}} \times \int_S I_\nu(s\frac{R}{c})\frac{\mathbf{f}_k(\mathbf{r}') \times \mathbf{R}}{R^2}\mathrm{d}S'\mathrm{d}S \tag{3.80}$$

$$d_{mk,\nu} = \frac{1}{4\pi}\int_S \mathbf{f}_m(\mathbf{r}) \cdot \hat{\mathbf{n}} \times \int_S I_\nu(s\frac{R}{c})\frac{\mathbf{f}_k(\mathbf{r}') \times \mathbf{R}}{R^3}\mathrm{d}S'\mathrm{d}S \tag{3.81}$$

$$\begin{aligned}
f_{mk,i} &= \frac{1}{2}\int_S \mathbf{f}_m(\mathbf{r}) \cdot \mathbf{f}_k(\mathbf{r}')\mathrm{d}\mathbf{S} \\
e_{m,i} &= \int_S \mathbf{f}_m(\mathbf{r}) \cdot \int_0^\infty \phi_i(st)\underline{\mathbf{E}}^i(\mathbf{r},t)\mathrm{d}(st)\mathrm{d}S \\
\dot{e}_{m,i} &= \int_S \mathbf{f}_m(\mathbf{r}) \cdot \int_0^\infty \phi_i(st)\frac{\partial \underline{\mathbf{E}}^i(\mathbf{r},t)}{\partial t}\mathrm{d}(st)\mathrm{d}S \\
h_{m,i} &= \int_S \mathbf{f}_m(\mathbf{r}) \cdot \int_0^\infty \phi_i(st)\hat{\mathbf{n}} \times \mathbf{H}^i(\mathbf{r},t)\mathrm{d}(st)\mathrm{d}S
\end{aligned}$$

$$\dot{a}_{mk,\nu} = s\ a_{mk,\nu} \tag{3.82}$$

$$\dot{b}_{mk,\nu} = \frac{s}{2}\ b_{mk,\nu}. \tag{3.83}$$

Having assumed that all the lower orders of the expansion coefficients up to $i-1$ on the right-hand sides of (3.69)-(3.71) are known, the retained coefficients associated with the present order $j = i$ on the left sides are found for $i = 0, 1, 2, \ldots, N$ sequentially. The alternative approach in which the spatial testing is preformed before the temporal testing (as Section 3.4.2) results in the same final matrix equations, i.e. (3.69)-(3.71) with (3.78)-(3.83) coefficients instead. In ultimate simplifications of the both alternative cases, one can avoid approximating $\frac{R}{c}$ by the electric distance between the center of subdomains (3.65) to pull the retarded term $I_\nu(s\frac{R}{c})$ out of the space integrals as Section 4.2. Preventing this unrealistic assumption of no changes for the unknown transient quantity within the subdomains of (3.78)-(3.83) improves the accuracy of the conventional MOD methods in Section 4.2 and accomplishes what is called the advanced MOD (AMOD) recipes. That is to say all the MOD algorithms in the previous section can also generate AMOD schemes, when one prevents the approximation (3.65) to take I_ν out of the surface integrals.

3.4.4 Summation Reduction Technique

The dominant cost of the (A)MOD methods are mainly the computation of the left-hand side of (3.69)-(3.71) involving space-time convolution of the past solution samples with the potential terms (3.78)-(3.81). The computational cost of the (A)MOD schemes scales as $\mathcal{O}(M^2N^2)$. This statement, however, is not immediately true for the plain implementation of the classical MOD in the previous Sections. In fact, distinct calculation of the double temporal (order) summations $\sum_{j=1}^{i}\sum_{\iota=0}^{j-1}$ in (3.69)-(3.71) for every iteration $i = 0, 1, 2, \cdots, N$ rises the total CPU cycles to $\mathcal{O}(M^2N^3)$. Storing the summation of current vectors up to the latest solution to avoid the recalculation of $\sum_\iota$ contents causes additional memory overhead of $\mathcal{O}(MN^2)$. The temporal accumulation of the retarded potential interactions in $\sum_{\iota=0}^{j-1}$ relating to (3.48), (3.49), and (3.50) can be extracted from the convolution products by the zero-padded extensions of $[1\ 1\ \cdots]_{1\times N}$, $[1\ 2\ 3\ \cdots]_{1\times N}$, and

$[-1\ 1\ -1\ 1\ \cdots]_{1\times N}$ with length $2N+1$, respectively. These convolutions can be calculated via element-by-element multiplication in spectral domain so as to be combined with the (space-)time FFT in Section 4.4. More efficient alternative computational schemes, however, can be obtained by eliminating the innermost loops $\sum_{\iota}$, considering that the double temporal sums $\sum_{j=0}^{i}\sum_{\iota=0}^{j-1}$ roll up their previous value together with the already evaluated recent contributions in the associated single temporal summation terms $\sum_{i=1}^{j-1}$. Therefore, the efficient counterparts of (3.69)-(3.71) are

$$\sum_{k=1}^{M}\left[\frac{a_{mk,0}}{2}+b_{mk,0}\right]c_{k,i}=e_{m,i}-\left(\frac{\underline{a}_m^i}{2}+\frac{\underline{b}_m^i}{2}\right) - \left(\bar{a}_m^i-\bar{b}_m^i\right) \tag{3.84}$$

$$\sum_{k=1}^{M}\left[\frac{\dot{a}_{mk,0}}{4}+\dot{b}_{mk,0}\right]c_{k,i}=\dot{e}_{m,i}-\left(\frac{\underline{\dot{a}}_m^i}{4}+\underline{\dot{b}}_m^i\right) - \left(\bar{\dot{a}}_m^i+\bar{\ddot{a}}_m^i\right) \tag{3.85}$$

$$\sum_{k=1}^{M}\left[f_{mk}-\left(\frac{c_{mk,0}}{2}+d_{mk,0}\right)\right]c_{k,i}=h_{m,i}+\left(\frac{\underline{c}_m^i}{2}+\underline{d}_m^i\right) +\bar{c}_m^i \qquad m=1,2,\ldots,M \qquad i=0,1,2,\ldots,N \tag{3.86}$$

where the vectors

$$\underline{g}_m^i=\sum_{k=1}^{M}\sum_{j=0}^{i-1}g_{mk,\nu}c_{k,j} \quad g=a,b,\dot{a},\dot{b},c,d \tag{3.87}$$

and the rest can be generated recursively through only two real vector additions at every iteration

$$\bar{g}_m^i=\mathsf{b}\bar{g}_m^{i-1}+\underline{g}_m^{i-1}+\sum_{k=1}^{M}g_{mk,0}\,c_{k,i-1} \quad g=a,b,\dot{a},c \tag{3.88}$$

in which $\bar{g}_m^0=\underline{g}_m^0=0$ and the sign $\mathsf{b}=-1$ for $g=b$ and 1 otherwise. Additionally, for the DEFIE case

$$\bar{\ddot{a}}_m^i=\bar{\ddot{a}}_m^{i-1}+\bar{\dot{a}}_m^{i-1} \tag{3.89}$$

with $\bar{\ddot{a}}_m^0=0$. The space vectors $\bar{g}_m$ and $\underline{g}_m$ are filled on the fly, i.e. the two vectors are updated for every order i to preserve the overall storage requirements to $\mathcal{O}(NM^2)$. The relations (3.84)-(3.86) scale down the total computational burden of the (A)MOD methods from $\mathcal{O}(M^2N^3)$ operations in construction and solution of (3.69)-(3.71) to $\mathcal{O}(M^2N^2)$, comparable with that of the MOT schemes [73]. It is also worth mentioning that (3.84) facilitates direct use of the surface electric current density, not the Hertz potential, as the unknown variable to the extent that for the first time the computational efficiency of solving the original form of the EFIE becomes comparable with solving the DEFIE, what may never comes up in the MOT framework [73].

3.4.5 Alternative AMOD with Reduced Sums

Considering the solution of MFIE (3.71) or the magnetic vector potential ($a_{mk,\nu}$ term) in the EFIE (3.69), the innermost summation is related to the time derivative and it appears due to

$$\frac{\mathrm{d}}{\mathrm{d}t}L_j(st) = -\sum_{\iota=0}^{j-1} L_\iota(st).$$

Insertion of the exponential damping factor in the above relation gives

$$\frac{\mathrm{d}}{\mathrm{d}t}\phi_j(st) = -s\left(\frac{1}{2}\phi_j(st) + \sum_{\iota=0}^{j-1}\phi_\iota(st)\right). \tag{3.90}$$

Instead of using (3.48), thus, one may directly expand the derivative of the current $\frac{\partial \mathbf{J}(\mathbf{r},t)}{\partial t}$ in the MFIE (2.24) by

$$\frac{\mathrm{d}}{\mathrm{d}t}c_k(t) = -s\sum_{j=0}^{\infty} c_{k,j}\left(\frac{1}{2}\phi_j(st) + \sum_{\iota=0}^{j-1}\phi_\iota(st)\right). \tag{3.91}$$

This useful alternative expression for (3.48) has been disregarded in all relevant literature, except [74]. Applying the time testing, analogous to the stage (3.77), one obtains

$$\sum_{k=1}^{M}\sum_{j=0}^{i}\frac{1}{2}c_{k,j}\delta_{ij}\mathbf{f}_k(\mathbf{r}') - \hat{\mathbf{n}} \times \frac{1}{4\pi}\left[\frac{s}{c}\sum_{k=1}^{M}\sum_{j=0}^{i} c_{k,j}\int_{S_0}\mathcal{I}_\nu(s\frac{R}{c})\frac{R}{c}\frac{\mathbf{f}_k(\mathbf{r}')\times\mathbf{R}}{R^2}\mathrm{d}S'\right.$$
$$\left.+\sum_{k=1}^{M}\sum_{j=0}^{i} c_{k,j}\int_{S_0} I_\nu(s\frac{R}{c})\frac{\mathbf{f}_k(\mathbf{r}')\times\mathbf{R}}{R^3}\mathrm{d}S'\right] = \int_0^\infty \phi_i(st)\hat{\mathbf{n}}\times\mathbf{H}^i(\mathbf{r},t)\mathrm{d}(st) \tag{3.92}$$

with such a single derivative counterpart for (3.64)

$$\mathcal{I}_\nu(s\frac{R}{c}) = \frac{1}{s}\int_0^\infty \phi_i(st)\frac{\mathrm{d}}{\mathrm{d}t}\phi_j(st - s\frac{R}{c})\mathrm{d}(st) = \frac{1}{2}\left(\phi_\nu(s\frac{R}{c}) + \phi_{\nu-1}(s\frac{R}{c})\right). \tag{3.93}$$

Alternatively, passing the temporal sum in the big parentheses of (3.77) to the space integral also afford the simplification

$$\sum_{j=0}^{i}\left(\frac{1}{2}c_{k,j} + \sum_{\iota=0}^{j-1} c_{k,\iota}\right) I_\nu(s\frac{R}{c}) = \sum_{j=0}^{i} c_{k,j}\mathcal{I}_\nu(s\frac{R}{c}). \tag{3.94}$$

By virtue of either (3.93) or (3.94), the last term in (3.71) vanishes, i.e.,

$$\sum_{k=1}^{M}\left[f_{mk} - (c_{mk,0} + d_{mk,0})\right]c_{k,i} = h_{m,i} + \sum_{k=1}^{M}\sum_{j=0}^{i-1}(c_{mk,\nu} + d_{mk,\nu})\,c_{k,j} \tag{3.95}$$

where only I_ν in (3.80) has to be interchanged with $\mathcal{I}_\nu$ for defining the new $c_{mn,\nu}$ in (3.95). Similarly, when one replaces the current derivative in the EFIE (2.10) by (3.91) or handing the first bracket in (3.75) over to the integral and using (3.94), the term $\sum_\iota$ over $a_{mk,\nu}$

disappears in (3.69), once I_ν in (3.78) is replaced by $\mathcal{I}_\nu$. Using the temporal integration of the weighted Laguerre polynomials,

$$\int_0^t c_k(\tau)\mathrm{d}\tau = -\frac{2}{s}\sum_{j=0}^{\infty} c_{k,j}\left[\phi_j(st) + 2\sum_{\iota=0}^{j-1}(-1)^{j+\iota}\phi_\iota(st)\right],$$

instead of the transient coefficients' closed-form integral (3.50), returns the similar associated term in (3.69). For the DEFIE, also using

$$\frac{\partial^2}{\partial t^2}c_k(\tau)\mathrm{d}\tau = -s^2\sum_{j=0}^{\infty} c_{k,j}\left[\frac{1}{4}\phi_j(st) + \sum_{\iota=0}^{j-1}(j-\iota)\phi_\iota(st)\right],$$

instead of (3.49), similarly leads into (3.70). However, one can exploit the time derivative of (3.93)

$$\frac{\mathrm{d}}{\mathrm{d}t}\mathcal{I}_\nu(st) = \frac{1}{2}\frac{\mathrm{d}}{\mathrm{d}t}\left[\phi_\nu(st) + \phi_{\nu-1}(st)\right]$$

and define

$$\dot{\mathcal{I}}_\nu(s\frac{R}{c}) = \frac{1}{4}\phi_\nu(s\frac{R}{c}) + \frac{3}{4}\phi_{\nu-1}(s\frac{R}{c}) + \sum_{\iota=0}^{\nu-2}\phi_\iota(s\frac{R}{c}) \tag{3.96}$$

in which the summation is bound up to ν not $j-1$ anymore. This yields a new equivalent formulation for (3.76)

$$\frac{\mu s^2}{4\pi}\sum_{k=1}^{M}\sum_{j=0}^{i} c_{k,j}\int_S \frac{\dot{\mathcal{I}}_\nu(s\frac{R}{c})\mathbf{f}_k(\mathbf{r}')}{R}\mathrm{d}S' - \frac{1}{4\pi\epsilon}\sum_{k=1}^{M}\sum_{j=0}^{i} c_{k,j}\int_S \nabla_{\mathbf{r}}\left[\frac{I_\nu(s\frac{R}{c})\nabla_{\mathbf{r}'}\cdot\mathbf{f}_k(\mathbf{r}')}{R}\right]\mathrm{d}S'$$
$$= \int_0^\infty \phi_i(st)\frac{\partial\underline{\mathbf{E}}^i(\mathbf{r},t)}{\partial t}\mathrm{d}(st) \tag{3.97}$$

that ends up with

$$\sum_{k=1}^{M}\left[\dot{a}_{mk,0} + \dot{b}_{mk,0}\right]c_{k,i} = \dot{e}_{m,i} - \sum_{k=1}^{M}\sum_{j=0}^{i-1}\left[\dot{a}_{mk,\nu} + \dot{b}_{mk,\nu}\right]c_{k,j} \tag{3.98}$$

where only I_ν in (3.82) has to be interchanged with $\dot{\mathcal{I}}_\nu$ for the definition of new $\dot{a}_{mn,l}$ in (3.98).

The interior summations in the conventional closed-form expressions for the derivations and integration of the Laguerre expansion (3.48)-(3.50) can be removed using specific linear combinations of future Laguerre polynomial order(s) in the initial expansion. For instance, to obtain a simple analytic formulation for the second derivation, more compact than (3.48) and (3.96), instead of classical inception (3.46), one may initially represent the unknown current in (3.51) by the superposition of every three orthogonal weighted Laguerre polynomials [75] (Fig. 3.6), that is

$$\mathbf{J}(\mathbf{r},t) = \sum_{k=1}^{M}\sum_{j=0}^{\infty} c_{k,j}\left[\phi_j(st) - 2\phi_{j+1}(st) + \phi_{j+2}(st)\right]\mathbf{f}_k(\mathbf{r}). \tag{3.99}$$

This allows the double derivative on the vector potential in (2.11) to be replaced by

$$\frac{\partial^2}{\partial t^2}\mathbf{J}(\mathbf{r},t)=\sum_{k=1}^{M}\sum_{j=0}^{\infty}c_{k,j}\left[\phi_j(st)+2\phi_{j+1}(st)+\phi_{j+2}(st)\right]\mathbf{f}_k(\mathbf{r}). \tag{3.100}$$

Performing the Petrov-Galerkin time testing by $<\phi_i(st),\cdot>$ on the DEFIE (2.11) gives

$$\frac{\mu s^2}{2\pi}\sum_{k=1}^{M}\sum_{j=0}^{i}c_{k,j}\int_S\left[\mathcal{I}_\nu(s\frac{R}{c})-\mathcal{I}_{\nu-2}(s\frac{R}{c})\right]\frac{\mathbf{f}_n(\mathbf{r}')}{R}\mathrm{d}S'$$
$$-\frac{1}{4\pi\epsilon}\sum_{k=1}^{M}\sum_{j=0}^{i}c_{k,j}\int_S\nabla_{\mathbf{r}}\left[I_\nu^{\mathrm{new}}(s\frac{R}{c})\frac{\nabla_{\mathbf{r}'}\cdot\mathbf{f}_n(\mathbf{r}')}{R}\right]\mathrm{d}S'=\int_0^\infty\phi_i(st)\frac{\partial\underline{\mathbf{E}}^i(\mathbf{r},t)}{\partial t}\mathrm{d}(st) \tag{3.101}$$

where

$$I_\nu^{\mathrm{new}}(s\frac{R}{c})=\left[I_\nu(s\frac{R}{c})-2I_{\nu-1}(s\frac{R}{c})+I_{\nu-2}(s\frac{R}{c})\right].$$

Applying the space testing to (3.101) results in a final matrix equation equivalent to (3.98) in which $\sum_\iota$ has been vanished.

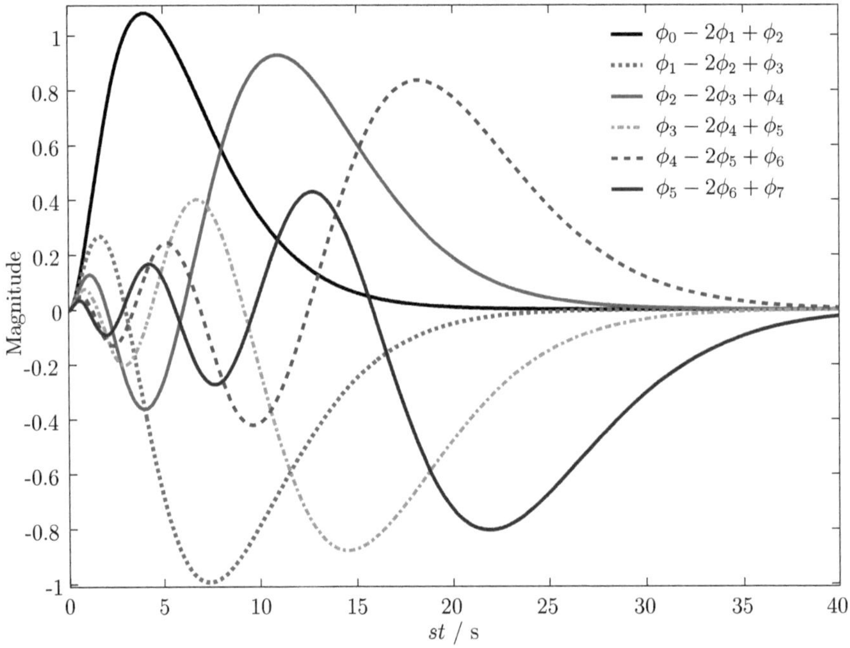

Figure 3.6: New entire-domain orthogonal time basis functions obtained by the combination of every three orthogonal orders of the weighted Laguerre polynomials.

Eventually for all the introduced schemes, such a following matrix equation is constituted and solved for all $m = 1, 2, \ldots, M$ at every i recursively

$$\bar{\bar{Z}}_0 \bar{I}_i = \bar{V}_i - \bar{\gamma}_{i-1} - \sum_{j=0}^{i-1} \bar{\bar{Z}}_\nu \bar{I}_j \qquad i = 0, 1, 2, \ldots, N, \tag{3.102}$$

where the $\sum_j$ represents that in (3.87) for the algorithms (3.84)-(3.86) and $\bar{\gamma}_{i-1} = 0$ for the schemes (3.95), (3.98), (3.101), or the FDDM (3.130), Fig. 3.7. Comparing (3.102) with the recursive final equations for the MOT method (3.19) better clarifies that the progression on polynomial orders i is tantamount to marching on time instances n and the role of order differences ν resembles to the contribution of the retarded samples $n - r$. Since $\bar{I}_0 \neq 0$ for the (A)MOD or CQM, we let $\bar{\bar{Z}}_i \bar{I}_0$ in (3.102) to be multiplied conventionally, so as to introduce FFT accelerated ways for the computation of $\sum_{j=1}^{i-1}$ in (3.102) or $\sum_{r=1}^{n-1}$ in (3.19) on a common platform in Section 4.4.

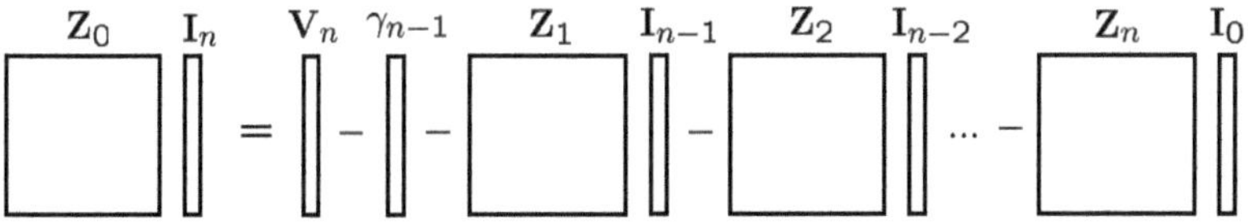

Figure 3.7: The recall of all already-calculated impedance matrices and known current vectors to construct the right side of the matrix equation in (A)MOD or FDDM.

3.4.6 Marching-on-in-Hermite Polynomials

To numerically solve the TDIE (2.10), (2.11), or (2.24) instead of marching on Laguerre functions, another choice of entire-domain temporal basis functions for the transient coefficients $c_k(t)$ in (3.46) are associated Hermite functions $h_j(st)$ [76]

$$c_k(t) = \sum_{j=0}^{\infty} c_{k,j} h_j(st)$$

where the weighted Hermite polynomials $h_j(st) = \frac{1}{\sqrt{j! 2^j \sqrt{\pi}}} e^{-(st)^2/2} H_j(st)$ [77]. The scaling factor s controls the temporal support provided by the expansion. The larger the scaling factor is, the finer the resolution in time is. The Hermite-Rodriguez polynomial of order j, $H_j(st)$, can be recursively regenerated in a stable manner from their lower orders

$$\begin{aligned} H_0(t) &= 1, \;\; H_1(t) = 2t \\ H_j(t) &= 2tH_{j-1}(t) - 2(j-1)H_{j-2}(t) \qquad \text{for } j \geq 2,\, t \geq 0. \end{aligned}$$

The recursion relation can also be expressed as

$$h_j(st) = \frac{1}{\sqrt{j}} \left[\sqrt{2}\, t h_{j-1}(st) - \sqrt{j-1} h_{j-2}(st) \right] \qquad j \geq 2. \tag{3.103}$$

An advantage feature of the Hermite polynomials is that they are orthogonal with respect

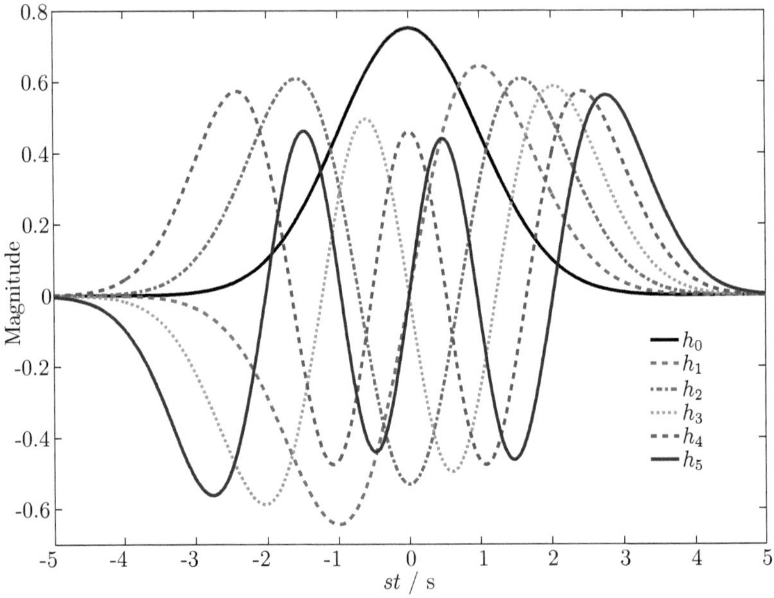

Figure 3.8: Hermite functions (Weighted Hermite polynomials) of different orders.

to the weighting function e^{-t^2}, that is

$$\int_{-\infty}^{\infty} e^{-t^2} H_i(t) H_j(t) \mathrm{d}t = \delta_{ij} 2^j j! \sqrt{\pi} \tag{3.104}$$

or equivalently for the Hermite functions

$$\int_{-\infty}^{\infty} h_i(st) h_j(st) \mathrm{d}(st) = \delta_{ij}$$

Therefore, $h_j(st)$ $j = 0, 1, 2, \ldots$ provide a complete orthogonal set of basis functions in $L^2(R)$ that decay to zero as $t \to \infty$, as exhibited in Fig. 3.8. The scaling factor s controls the temporal support provided by the expansion. The larger the scaling factor is, the finer the resolution in time is. The closed-form analytical expressions for the first and the second derivatives as well as the indefinite integral of the Hermite transform coefficients with respect to time are available, e.g. using the Appell sequence $\frac{\mathrm{d}}{\mathrm{d}t} H_j(t) = 2j H_{j-1}(t)$

$$\frac{\mathrm{d}}{\mathrm{d}t} h_j(st) = s \left[\sqrt{\frac{j}{2}} h_{j-1}(st) - \sqrt{\frac{j+1}{2}} h_{j+1}(st) \right]$$

$$\frac{\mathrm{d}}{\mathrm{d}t} h_j(st) = s \left[\sqrt{2j}\, h_{j-1}(st) - \sqrt{2}\, st\, h_j(st) \right] \qquad n \geq 1 \tag{3.105}$$

$$\frac{\mathrm{d}}{\mathrm{d}t} h_0(st) = -st\, h_0(st) \qquad n = 0 \tag{3.106}$$

$$\begin{aligned}
\frac{\mathrm{d}}{\mathrm{d}t}c_k(t) &= s\sum_{j=0}^{\infty}\left(\sqrt{\frac{j}{2}}c_{k,j-1}+\sqrt{\frac{j+1}{2}}c_{k,j+1}\right)h_j(st) \qquad (3.107)\\
\frac{d^2}{\mathrm{d}t^2}h_j(st) &= \frac{s^2}{2}\left[\sqrt{j(j-1)}h_{j-2}(st)+h_j(st)-\sqrt{(j+1)(j+2)}h_{j+2}(st)\right]\\
\frac{d^2}{\mathrm{d}t^2}c_k(t) &= \frac{s^2}{2}\sum_{j=0}^{\infty}\left[\sqrt{j(j-1)}c_{k,j-2}+c_{k,j}+\sqrt{(j+1)(j+2)}c_{k,j+2}\right]h_j(st) \qquad (3.108)
\end{aligned}$$

The Galerkin's time testing formulation for the Hermite functions is given in Appendix 8.5.

3.5 Finite Difference Delay Modeling (FDDM)

The temporal discretization of the time-domain integral equations (TDIE) is commonly accomplished by either the implicit marching-on-in-time (MOT) schemes using subdomain Lagrange polynomial interpolation (Section 3.1) or the always-stable marching-on-in-order/degrees (MOD) of Laguerre entire-domain bases (Section 3.4). An alternative approach for discretizing the time convolution integrals in the TDIE, competitive to the time basis functions expansions in the MOT or MOD recipes, is the Lubich's convolution quadrature methods (CQM) [78], using the (first or) second order backward finite difference (BFD) approximations in the Laplace domain. The underlying physics describing the wave scattering process is time invariant, as the material properties do not change over time. The CQM are utilized to transform continuous-time representation of the time-invariant integral kernel (system transfer function) to discrete-time domain while approximating the TDIE derivatives in the spectral domain. The CQM are called finite difference delay modeling (FDDM) when the scattering analysis of arbitrarily shaped three-dimensional (3D) structures is carried out in a marching style [79]. The FDDM is a provably stable method when the Lubich's convolution quadrature method for the time discretization is used in conjunction with the Galerkin moment method for the spatial discretization.

In the FDDM method, a conformal mapping from the Laplace domain to the z-transform domain (bilinear transform) based on a finite difference formula accomplishes the discretization in the z-domain, and the result is inverse transformed to create a time-domain method. The temporal discretization is carried out by either first or second order BFD approximation when the transformed TDIE is mapped from the Laplace domain to the z-transform domain. In the mathematical literature, the method is called convolution quadrature when it applies to the single layer potential for the Helmholtz operator [80]. It is also called Tustin's method in digital signal processing and control theory to transform continuous-time representation (transfer function) of a linear time-invariant system to discrete-time domain. The bilinear transformation preserves the stability by exact mapping every point of the $j\omega$-axis in the s-plane onto the unit circle $|z| = 1$ in the z-plane. Following this approach, each frequency response of the continuous-time system can be processed using a discrete-time filtering technique. The numerical solution of retarded functional equations can benefit from this frequency wrapping once the system's unit delays are replaced by first order all-pass filters z^{-1} in the discrete domain, as schematically depicted in Fig. 3.9.

Weile *et al.* [79] determine the time derivative of the CFIE in conjunction with the higher-order spatial bases. We adopt the FDDM also to the original forms of the EFIE

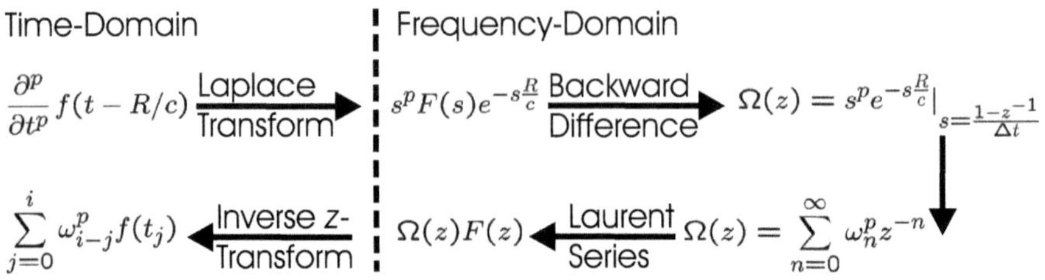

Figure 3.9: Block diagram of the time discretization/integration procedures by the CQM or FDDM methods.

containing a time integral as well as the MFIE with a single time derivative, using the linearly-varying divergence-conforming space basis functions. To this end, the time-domain EFIE (2.10), DEFIE (2.11), and MFIE (2.24) are respectively transferred to the Laplace domain

$$\frac{\mu s}{4\pi}\int_S \tilde{\mathbf{J}}(\mathbf{r}',s)\frac{e^{-s\frac{R}{c}}}{R}\mathrm{d}S' - \frac{\nabla_{\mathbf{r}}}{4\pi\epsilon s}\int_S \nabla_{\mathbf{r}'}.\tilde{\mathbf{J}}(\mathbf{r}',s)\frac{e^{-s\frac{R}{c}}}{R}\mathrm{d}S'\bigg|_{\tan} = \tilde{\underline{\mathbf{E}}}^i(\mathbf{r},s) \tag{3.109}$$

$$\frac{\mu s^2}{4\pi}\int_S \tilde{\mathbf{J}}(\mathbf{r}',s)\frac{e^{-s\frac{R}{c}}}{R}\mathrm{d}S' - \frac{\nabla_{\mathbf{r}}}{4\pi\epsilon}\int_S \nabla_{\mathbf{r}'}.\tilde{\mathbf{J}}(\mathbf{r}',s)\frac{e^{-s\frac{R}{c}}}{R}\mathrm{d}S'\bigg|_{\tan} = s\tilde{\underline{\mathbf{E}}}^i(\mathbf{r},s) \tag{3.110}$$

$$\frac{\tilde{\mathbf{J}}(\mathbf{r},s)}{2} - \frac{1}{4\pi}\hat{\mathbf{n}}\times\int_{S_0}\left[\frac{s}{c}\tilde{\mathbf{J}}(\mathbf{r}',s)\times\frac{\mathbf{R}}{R^2} + \tilde{\mathbf{J}}(\mathbf{r}',s)\times\frac{\mathbf{R}}{R^3}\right]e^{-s\frac{R}{c}}\mathrm{d}S' = \hat{\mathbf{n}}\times\tilde{\mathbf{H}}^i(\mathbf{r},s). \tag{3.111}$$

The unknown induced surface current density $\tilde{\mathbf{J}}(\mathbf{r},s)$ is approximately expanded using Bernoulli's separation of variables in the space and Laplace domains by

$$\tilde{\mathbf{J}}(\mathbf{r},s) = \sum_{k=1}^{M}\tilde{I}_k(s)\mathbf{f}_k(\mathbf{r}). \tag{3.112}$$

where $\tilde{I}_k(s)$ are unknown weighting coefficients of the vector spatial basis functions $\mathbf{f}_k(\mathbf{r})$. Substituting (3.112) in the TDIE (3.109), (3.110), and (3.111) and performing the Galerkin's testing procedure in space using the same set of vector basis functions $\mathbf{f}_m(\mathbf{r})$, $m = 1,2,\ldots,M$, respectively, give

$$\frac{\mu s}{4\pi}\sum_{k=1}^{M}\tilde{I}_k(s)\int_S \mathbf{f}_m(\mathbf{r})\cdot\int_S\frac{\mathbf{f}_k(\mathbf{r}')}{R}e^{-s\frac{R}{c}}\mathrm{d}S'\mathrm{d}S$$
$$+\frac{1}{4\pi\epsilon s}\sum_{k=1}^{M}\tilde{I}_k(s)\int_S\nabla_{\mathbf{r}}\cdot\mathbf{f}_m(\mathbf{r})\int_S\frac{\nabla_{\mathbf{r}'}\cdot\mathbf{f}_k(\mathbf{r}')}{R}e^{-s\frac{R}{c}}\mathrm{d}S'\mathrm{d}S = \int_S\mathbf{f}_m(\mathbf{r})\cdot\tilde{\underline{\mathbf{E}}}^i(\mathbf{r},s)\mathrm{d}S \tag{3.113}$$

$$\frac{\mu s^2}{4\pi}\sum_{k=1}^{M}\tilde{I}_k(s)\int_S \mathbf{f}_m(\mathbf{r})\cdot\int_S\frac{\mathbf{f}_k(\mathbf{r}')}{R}e^{-s\frac{R}{c}}\mathrm{d}S'\mathrm{d}S$$
$$+\frac{1}{4\pi\epsilon}\sum_{k=1}^{M}\tilde{I}_k(s)\int_S\nabla_{\mathbf{r}}\cdot\mathbf{f}_m(\mathbf{r})\int_S\frac{\nabla_{\mathbf{r}'}\cdot\mathbf{f}_k(\mathbf{r}')}{R}e^{-s\frac{R}{c}}\mathrm{d}S'\mathrm{d}S = s\int_S\mathbf{f}_m(\mathbf{r})\cdot\tilde{\underline{\mathbf{E}}}^i(\mathbf{r},s)\mathrm{d}S \tag{3.114}$$

$$\frac{1}{2}\sum_{k=1}^{M}\tilde{I}_k(s)\int_S \mathbf{f}_m(\mathbf{r})\cdot\mathbf{f}_k(\mathbf{r}')\mathrm{d}S - \frac{1}{4\pi}\left[\sum_{k=1}^{M}\frac{s}{c}\tilde{I}_k(s)\int_S \mathbf{f}_m(\mathbf{r})\cdot\hat{\mathbf{n}}\times\int_S\frac{\mathbf{f}_k(\mathbf{r}')\times\mathbf{R}}{R^2}e^{-s\frac{R}{c}}\mathrm{d}S'\mathrm{d}S\right.$$
$$\left.+\sum_{k=1}^{M}\tilde{I}_k(s)\int_S \mathbf{f}_m(\mathbf{r})\cdot\hat{\mathbf{n}}\times\int_S\frac{\mathbf{f}_k(\mathbf{r}')\times\mathbf{R}}{R^3}e^{-s\frac{R}{c}}\mathrm{d}S'\mathrm{d}S\right] = \int_S \mathbf{f}_m(\mathbf{r})\cdot\hat{\mathbf{n}}\times\tilde{\mathbf{H}}^i(\mathbf{r},s)\mathrm{d}S. \quad (3.115)$$

Each of the above three equations forms such a matrix equation in Laplace domain

$$\left[\tilde{\mathrm{Z}}(s)\right]_{mk}\left[\tilde{\mathrm{I}}(s)\right]_k = \left[\tilde{\mathrm{V}}(s)\right]_m \qquad m = 1, 2, \ldots, M. \quad (3.116)$$

In the absence of inhomogeneous initial conditions, multiplication by s in the Laplace domain corresponds to the temporal differentiation. Thus, one may replace s with a finite difference approximation[1]. Considering the unit delay property of the z-transform in Section 8.6, the first-order backward Euler and the second-order backward difference approximations to s are obtained respectively by

$$s = \frac{1 - z^{-1}}{\Delta t} \quad (3.117)$$

$$s = \frac{3 - 4z^{-1} + z^{-2}}{\Delta t}. \quad (3.118)$$

The impedance matrices $\tilde{\mathrm{Z}}(s)$ are function of $s^p e^{-s\frac{R}{c}}$, $p = -1, 0, 1, 2$. Substituting (3.117) and (3.118) to the functions $\tilde{\mathrm{Z}}(s)$,

$$s^p e^{-s\frac{R}{c}}\Big|_{s=\frac{1-z^{-1}}{\Delta t}} = \sum_{n=0}^{\infty}\omega_n^p z^{-n} \quad (3.119)$$

$$s^p e^{-s\frac{R}{c}}\Big|_{s=\frac{3-4z^{-1}+z^{-2}}{\Delta t}} = \sum_{n=0}^{\infty}\omega_n^p z^{-n}, \quad (3.120)$$

and assuming $\omega_n^p = 0$ for $n < 0$, the Laurent series expansion gives the time sequence ω_n^p versus $\xi = \frac{R}{c\Delta t}$ respectively for (3.117) and (3.118) substitutions

$$\omega_n^0 = \frac{\xi^n e^{-\xi}}{n!} \quad (3.121)$$

$$\omega_n^0 = \frac{1}{n!}\left(\frac{\xi}{2}\right)^{\frac{n}{2}} e^{-\frac{3}{2}\xi}H_n(\sqrt{2\xi}) \quad (3.122)$$

and for the higher order derivation $p > 0$ terms in (3.117) and (3.118) respectively [79, 80]

$$\omega_n^p = \frac{1}{\Delta t}\left(\omega_n^{p-1} - \omega_{n-1}^{p-1}\right) \qquad p > 0 \quad (3.123)$$

[1]BFD approximations of greater than second order are never absolutely stable.

$$\omega_n^p = \frac{1}{2\Delta t}\left(3\omega_n^{p-1} - 4\omega_{n-1}^{p-1} + \omega_{n-2}^{p-1}\right) \qquad p > 0. \tag{3.124}$$

Therefore, using the second-order backward difference formula to approximate the Laplace variable (differentiation operator) s, the inverse z-transform of the coefficient matrix appears as a function of ordered Hermite polynomials. For $p = -1$ [81],

$$\omega_n^{-1} = \Delta t\, e^{-\xi} \sum_{\iota=0}^{n} \frac{\xi^{n-\iota}}{(n-\iota)!} = \Delta t\, e^{-\xi} \sum_{\iota=0}^{n} \frac{\xi^{\iota}}{\iota!}. \tag{3.125}$$

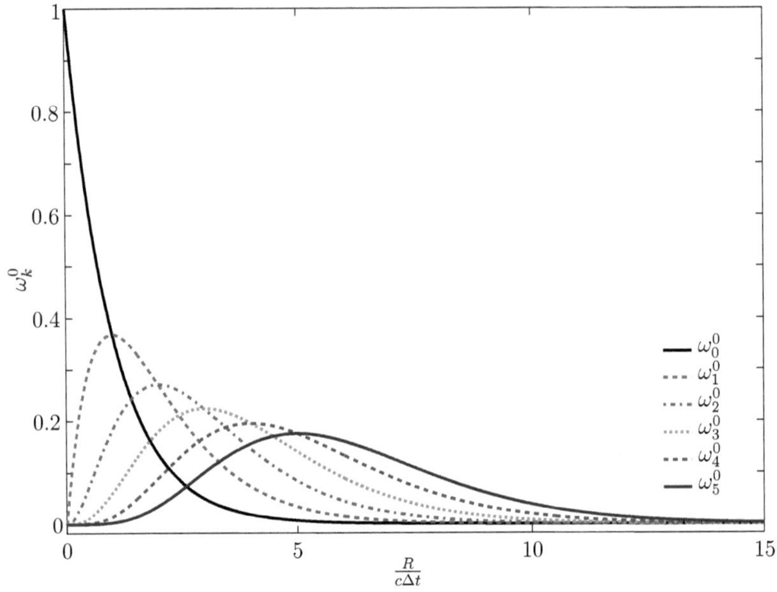

Figure 3.10: Coefficients ω_k^0 for the first-order backward-difference approximation as a function of the electric distance.

Taking the inverse z-transform (3.113), (3.114), and (3.115), the multiplications in the frequency-domain become convolutions in the time-domain [82],

$$\frac{\mu}{4\pi}\sum_{k=1}^{M}\sum_{j=0}^{i} I_k(t_j)\int_S \mathbf{f}_m(\mathbf{r})\cdot\int_S \omega_{i-j}^{1}\frac{\mathbf{f}_k(\mathbf{r}')}{R}\mathrm{d}S'\mathrm{d}S$$
$$+\frac{1}{4\pi\epsilon}\sum_{k=1}^{M}\sum_{j=0}^{i} I_k(t_j)\int_S \nabla_{\mathbf{r}}\cdot\mathbf{f}_m(\mathbf{r})\int_S \omega_{i-j}^{-1}\frac{\nabla_{\mathbf{r}'}\cdot\mathbf{f}_k(\mathbf{r}')}{R}\mathrm{d}S'\mathrm{d}S = \int_S \mathbf{f}_m(\mathbf{r})\cdot\underline{\mathbf{E}}^i(\mathbf{r},t_i)\mathrm{d}S \tag{3.126}$$

$$\frac{\mu}{4\pi}\sum_{k=1}^{M}\sum_{j=0}^{i} I_k(t_j)\int_S \mathbf{f}_m(\mathbf{r})\cdot\int_S \omega_{i-j}^{2}\frac{\mathbf{f}_k(\mathbf{r}')}{R}\mathrm{d}S'\mathrm{d}S$$
$$+\frac{1}{4\pi\epsilon}\sum_{k=1}^{M}\sum_{j=0}^{i} I_k(t_j)\int_S \nabla_{\mathbf{r}}\cdot\mathbf{f}_m(\mathbf{r})\int_S \omega_{i-j}^{0}\frac{\nabla_{\mathbf{r}'}\cdot\mathbf{f}_k(\mathbf{r}')}{R}\mathrm{d}S'\mathrm{d}S = s\int_S \mathbf{f}_m(\mathbf{r})\cdot\underline{\mathbf{E}}^i(\mathbf{r},t_i)\mathrm{d}S \tag{3.127}$$

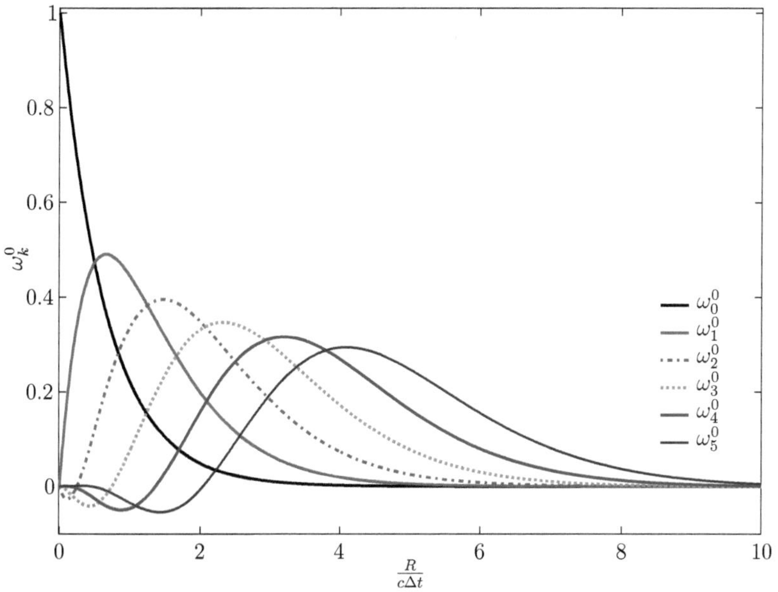

Figure 3.11: Coefficients ω_k^0 for the second-order backward-difference approximation as a function the electric distance.

$$\frac{1}{2}\sum_{k=1}^{M} I_k(t_i)\int_S \mathbf{f}_m(\mathbf{r})\cdot\mathbf{f}_k(\mathbf{r}')\mathrm{d}S - \frac{1}{4\pi}\left[\sum_{k=1}^{M}\sum_{j=0}^{i} I_k(t_j)\int_S \mathbf{f}_m(\mathbf{r})\cdot\hat{\mathbf{n}}\times\int_S \omega_{i-j}^{1}\frac{\mathbf{f}_k(\mathbf{r}')\times\mathbf{R}}{cR^2}\mathrm{d}S'\mathrm{d}S\right.$$
$$\left.+\sum_{k=1}^{M}\sum_{j=0}^{i} I_k(t_j)\int_S \mathbf{f}_m(\mathbf{r})\cdot\hat{\mathbf{n}}\times\int_S \omega_{i-j}^{0}\frac{\mathbf{f}_k(\mathbf{r}')\times\mathbf{R}}{R^3}\mathrm{d}S'\mathrm{d}S\right] = \int_S \mathbf{f}_m(\mathbf{r})\cdot\hat{\mathbf{n}}\times\mathbf{H}^i(\mathbf{r},t_i)\mathrm{d}S. \quad (3.128)$$

where $t_i = i\Delta t$, or in general when (3.116) is transferred back to the time-domain, the sequential calling of the coefficient matrices forms a block-triangular Toeplitz system matrix

$$\sum_{j=0}^{\infty}\bar{\bar{\mathrm{Z}}}_{i-j}\bar{\mathrm{I}}_j \equiv \sum_{j=0}^{i}\bar{\bar{\mathrm{Z}}}_{i-j}\bar{\mathrm{I}}_j = \bar{\mathrm{V}}_i \qquad 0 \le i < \infty. \quad (3.129)$$

Having assumed that all the coefficients up to $i-1$ are known, they are sent to the right-hand sides of (3.131) and the retained coefficients associated with the present weight $j = i$ are found for $i = 0, 1, 2, \ldots, N$ sequentially,

$$\bar{\bar{\mathrm{Z}}}_0\bar{\mathrm{I}}_i = \bar{\mathrm{V}}_i - \sum_{j=0}^{i-1}\bar{\bar{\mathrm{Z}}}_{i-j}\bar{\mathrm{I}}_j \qquad 0 \le i < N < \infty. \quad (3.130)$$

Using a proper cutoff strategy [82] before preforming the matrix-vector products, the high storage cost is reduced for marching on smooth tails of the late transient response, i.e.,

(3.130) becomes

$$\bar{\bar{\mathrm{Z}}}_0\bar{\mathrm{I}}_i = \bar{\mathrm{V}}_i - \sum_{j=\max(0,i-N_g)}^{i} \bar{\bar{\mathrm{Z}}}_{i-j}\bar{\mathrm{I}}_j, \qquad 0 \le i < N \tag{3.131}$$

where the almost-zero late-appearing matrices $\bar{\bar{\mathrm{Z}}}_n \simeq 0$, $n > N_g$ can be eliminated from the construction of the matrix equation, after marching on the solutions up to $\xi^n_{\max} << n!$ [83]. Thus, the knowledge of the largest electrical dimension of the structure is sufficient to determine after how long marching on time samples, the newly appearing interaction matrices can be neglected without disturbing the late-time accuracy and stability. Additionally, by using the inherently parallelizable time-FFT for the discrete convolutions as in Section 4.4, the computational complexity is reduced further. Note that the singularity may exist for several terms of $\omega^p_n \frac{1}{R}$, $n = 0, 1, 2, \ldots$ in (3.121)-(3.124) when $R \to 0$. For instance, for the treatment of singularity associated with $\omega^2_{i-j}\frac{1}{R}$ in (3.124) five terms $j = i, i-1, \ldots, i-4$ have to be handled separately.

3.5.1 Convolution Quadrature Methods (CQM)

Substituting (3.112) in the EFIE (2.10), DEFIE (2.11), and MFIE (2.24)

$$\frac{\mu s}{4\pi}\sum_{k=1}^{M}\tilde{I}_k(s)\int_S \frac{\mathbf{f}_k(\mathbf{r}')}{R}e^{-s\frac{R}{c}}\mathrm{d}S' - \frac{\nabla_{\mathbf{r}}}{4\pi\epsilon s}\sum_{k=1}^{M}\tilde{I}_k(s)\int_S \frac{\nabla_{\mathbf{r}'}\cdot\mathbf{f}_k(\mathbf{r}')}{R}e^{-s\frac{R}{c}}\mathrm{d}S' = \tilde{\underline{\mathbf{E}}}^i(\mathbf{r}, s) \tag{3.132}$$

$$\frac{\mu s^2}{4\pi}\sum_{k=1}^{M}\tilde{I}_k(s)\int_S \frac{\mathbf{f}_k(\mathbf{r}')}{R}e^{-s\frac{R}{c}}\mathrm{d}S' - \frac{\nabla_{\mathbf{r}}}{4\pi\epsilon}\sum_{k=1}^{M}\tilde{I}_k(s)\int_S \frac{\nabla_{\mathbf{r}'}\cdot\mathbf{f}_k(\mathbf{r}')}{R}e^{-s\frac{R}{c}}\mathrm{d}S' = s\tilde{\underline{\mathbf{E}}}^i(\mathbf{r}, s) \tag{3.133}$$

$$\begin{aligned}\frac{1}{2}\sum_{k=1}^{M}\tilde{I}_k(s)\mathbf{f}_k(\mathbf{r}') - \frac{1}{4\pi}\hat{\mathbf{n}}\times\Bigg[\sum_{k=1}^{M}\frac{s}{c}\tilde{I}_k(s)\int_S \frac{\mathbf{f}_k(\mathbf{r}')\times\mathbf{R}}{R^2}e^{-s\frac{R}{c}}\mathrm{d}S' \\ +\sum_{k=1}^{M}\tilde{I}_k(s)\int_S \frac{\mathbf{f}_k(\mathbf{r}')\times\mathbf{R}}{R^3}e^{-s\frac{R}{c}}\mathrm{d}S'\Bigg] = \hat{\mathbf{n}}\times\tilde{\mathbf{H}}^i(\mathbf{r}, s).\end{aligned} \tag{3.134}$$

Taking unilateral inverse z-transform from the one-sided Laplace transform of the ideal sampled quantities in (3.132)-(3.134), one obtains convolutions with discrete time-domain samples

$$\begin{aligned}&\frac{\mu}{4\pi}\sum_{k=1}^{M}\sum_{j=0}^{i} I_k(t_j)\int_S \omega^1_{i-j}\frac{\mathbf{f}_k(\mathbf{r}')}{R}\mathrm{d}S' \\ &+\frac{\nabla_{\mathbf{r}}}{4\pi\epsilon}\sum_{k=1}^{M}\sum_{j=0}^{i} I_k(t_j)\int_S \omega^{-1}_{i-j}\frac{\nabla_{\mathbf{r}'}\cdot\mathbf{f}_k(\mathbf{r}')}{R}\mathrm{d}S' = \underline{\mathbf{E}}^i(\mathbf{r}, t_i)\end{aligned} \tag{3.135}$$

$$\begin{aligned}&\frac{\mu}{4\pi}\sum_{k=1}^{M}\sum_{j=0}^{i} I_k(t_n)\int_S \omega^2_n\frac{\mathbf{f}_k(\mathbf{r}')}{R}\mathrm{d}S' \\ &+\frac{\nabla_{\mathbf{r}}}{4\pi\epsilon}\sum_{k=1}^{M}\sum_{j=0}^{i} I_k(t_n)\int_S \omega^0_n\frac{\nabla_{\mathbf{r}'}\cdot\mathbf{f}_k(\mathbf{r}')}{R}\mathrm{d}S' = \frac{\partial}{\partial t}\underline{\mathbf{E}}^i(\mathbf{r}, t_i)\end{aligned} \tag{3.136}$$

$$\frac{1}{2}\sum_{k=1}^{M} I_k(t_i)\mathbf{f}_k(\mathbf{r}) - \frac{1}{4\pi}\hat{\mathbf{n}} \times \left[\sum_{k=1}^{M}\sum_{j=0}^{i} I_k(t_j)\int_S \omega^1_{i-j}\frac{\mathbf{f}_k(\mathbf{r}') \times \mathbf{R}}{cR^2}\mathrm{d}S'\right.$$
$$\left. + \sum_{k=1}^{M}\sum_{j=0}^{i} I_k(t_j)\int_S \omega^0_{i-j}\frac{\mathbf{f}_k(\mathbf{r}') \times \mathbf{R}}{R^3}\mathrm{d}S'\right] = \hat{\mathbf{n}} \times \mathbf{H}^i(\mathbf{r}, t_i). \quad (3.137)$$

Now, the Galerkin's testing procedure in space on (3.135)-(3.137) gives the final matrix equations (3.126)-(3.128), respectively. In dispersive dielectric material where the relative permittivity $\epsilon_r(s)$ and permeability $\mu_r(s)$ and the Green's functions are functions of s (e.g., in Debye equation for the complex permittivity), the CQM for time-discretization in the Laplace domain let the frequency-dependent characteristics be directly incorporated into the time-domain solver [84]. Implicit Runge-Kutta schemes can also be applied for the temporal discretization, i.e. mapping from the Laplace domain to z-domain. Thereafter, s parameter is replaced with a matrix function of z and the inverse z-transform is computed numerically using the discrete Fourier transform (DFT) [85]. The absolutely stable (A-stable) Radau IIA methods with two- and three-stage has third- and fifth-order convergence, respectively [85].

3.6 Symplectic Time Integration for Energy Conservation

The numerical accuracy and energy conservation are of great importance in real-life EM problems, such as tracing the influence of excited wake fields inside accelerator structures on moving charged particles over long time periods which is discussed in details in Section 6.11. In practice, the essential property needed for accurate time-domain field simulations is the fulfillment of the energy conservation law implied by Maxwell equations. The TDIE are commonly solved by the MOT, MOD, or FDDM methods. All these discretization schemes employ the Galerkin method in space whereas only the MOT applies the point-matching for time testing. The Green's function is a reciprocal function of observation and source points distance $|\mathbf{r}-\mathbf{r}'|$, and hence, the Galerkin method clearly preserves the symmetry of a scattering operator and inherently can satisfy conservation of energy [86]. Due to the use of the Galerkin technique at the core stage of the TDIE formulations, similar to the frequency-domain method of moments, the evaluation of mutual interactions of subdomains demands computation of double surface integrals (quadruple space quadratures). The inner integrals over source subdomains have been calculated either analytically or numerically whereas to lessen the matrix fill-in costs, the outer surface integrals has been widely replaced by the centroid value of the integrand or approximated by a few fixed collocation points. Consequently, the equivalent resulting lossless scattering matrices are asymmetric in which the prerequisite condition for the conservation of energy has been violated [86].

The energy dissipation due to numerical integration methods can lead to misleading results for large systems that need to be iterated for long time intervals. The solution is to use symplectic space-time integration methods that comply the Lorentz reciprocity theorem and fully conserve the energy. In this regard, the advantage of symmetric discretization in numerical solution of retarded functional integral equations is not so far clear. In the TDIE, the past solution samples contribute in determination of present status of the system, and hence, the solution accuracy is coupled to the late-time stability. Stability analysis of numerical methods is typically carried out based on the error bound of the approximate solution without considering the energy balance. To avoid the error propagation of asymmetric quadrature routines leading to unbounded energy, in this section the numerical calculation of the double surface integrals is studied in a totally symmetric way where the number of quadrature points are adaptively refined by simultaneous partitioning of source and observation subdomains. Probing the eigenvalue spectra of the retarded time dependent system reveals that many unwanted modes cause the propagation of numerical errors that may not even unveil the late-time instabilities due to considerations in spatial discretization. Among choices of the stable time integration methods, the one through which the EM energy is conserved are sought in the next section.

3.6.1 Symmetric Adaptive Refining Quadrature Routines

Semi-analytical formulations to evaluate the interior integrals is growingly used in the BEMs in which the symmetry of the discretization, the necessary condition for conservation properties, is still annihilated [87]. For fulfillment of energy conservation by the reciprocal Galerkin method in the frequency-domain MoM, the double surface integrals representing the mutual interaction of basis functions have to be numerically calculated in a fully symmetric way. This procedure causes better-conditioned coefficient matrix $\bar{\bar{\mathrm{Z}}}_0$, symmetric $\bar{\bar{Z}}_n$ for which only half of the matrix elements have to be calculated.

To perform numerical integrations, the vertices of the mutual triangles are first mapped to the points (0,0), (0,1), and (1,0) in the xy-plane and then a p-point quadrature rule is applied to the unit triangle. Of course, any constant multi-point quadrature rule can be considered at this stage, e.g. $p \in \{3, 4, 7, \ldots\}$ [64]. To adaptively control the precision of numerical integrations over the surface patches so as to guarantee a total quadrature error of less than an a priori set value ε, two-scale simultaneous refinements of the both patches are considered.

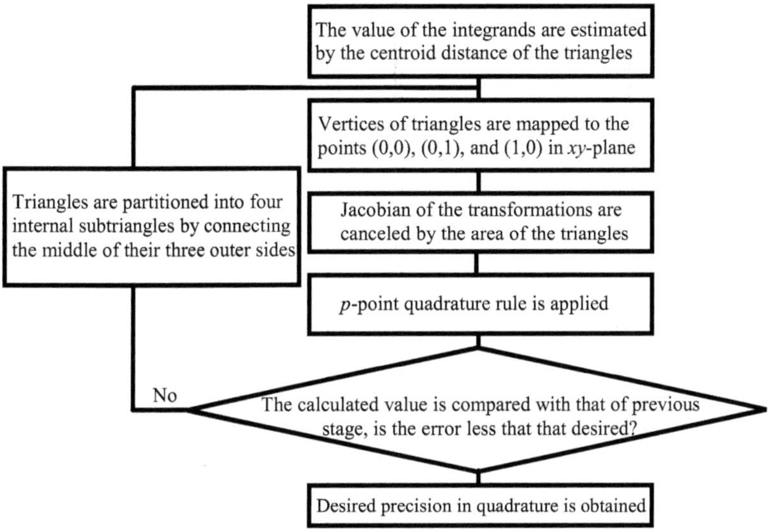

Figure 3.12: Adaptive numerical evaluation of the integrals for controling quadrature errors.

Fig. 3.13 and Fig. 3.14 illustrate how the number of quadrature points are adaptively refined by simultaneous partitioning of source and observation subdomains into four equilateral subcells through connecting the bisectors of every three or four sides. At the first stage, if the difference between the results of, for instance, a single-point and the p-point quadrature rule is greater than the prespecified error, the source and observation subdomains are partitioned preferably into four equilateral subtriangles by bisecting and connecting the middle of their three sides. Every half-scaled triangle is again mapped to the vertices of the unit triangle and the sum of their contributions is checked through comparison with the result of the previous stage. Subdivision of internal subtriangles and comparison of resultant quadrature values with those of the former stage is successively continued until a predefined precision ensuring sufficiently accurate results. Fig. 3.12 depicts the flowchart of the quadrature algorithm. Specifically, Fig. 3.13 shows a three-stage partitioning of coupled triangles.

The procedure is similar for quadrilateral subdomains (e.g., rooftop basis functions) except that at the beginning, the vertices of the corresponding quadrilateral are first mapped to the points (0,0), (0,1), (1,0), and (1,1) in the xy plane and then the four-point quadra-

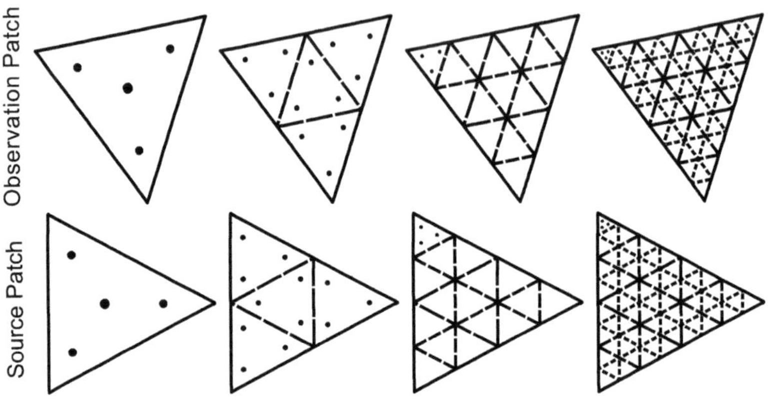

Figure 3.13: The early three stages in the symmetric subdivision of the coupled source and observation triangular subdomains for controlling the numerical error in calculation of the double surface integrals (quadruple space quadratures).

ture rule is applied to the unit squares. Apparently, the non-nodal nine-point quadrature rule ($p = 9$) [64] can also be employed at this stage instead.

As a more efficient extended case, one can also perform a higher order quadrature rule in parallel, and additionally compare the results of two quadrature routines before taking any decision for further partitioning. Note that the Jacobians of the transformation at every stage are a quarter fraction of the surface area of the underlying triangle, and since they are finally canceled during evaluation of (2.46)-(2.49), there is no need for explicit calculation and storage. In the singular cases, the double surface integrals are evaluated analytically [88, 89]. When the triangles T_m^p and T_k^q, respectively with the barycentric coordinates $(\lambda_1, \lambda_2, 1-\lambda_1-\lambda_2)$ and $(\lambda'_1, \lambda'_2, 1-\lambda'_1-\lambda'_2)$, are coincide with each other or share a common edge, for instance, since

$$\begin{aligned}\rho_m^p \cdot \rho'^q_k &= (\mathbf{r}-\mathbf{r}_m)\cdot(\mathbf{r}'-\mathbf{r}_k)\\ &= [(\mathbf{r}_1-\mathbf{r}_3)\lambda_1+(\mathbf{r}_2-\mathbf{r}_3)\lambda_2+\mathbf{r}_3-\mathbf{r}_m]\cdot[(\mathbf{r}_1-\mathbf{r}_3)\lambda'_1+(\mathbf{r}_2-\mathbf{r}_3)\lambda'_2+\mathbf{r}_3-\mathbf{r}_k],\end{aligned}$$

the vector integral (2.46) can be calculated by

$$\begin{aligned}\frac{1}{4}A_{mk}^{pq} &= \frac{1}{4{A_m^p}^2}\int_{T_m^p}\rho_m^p\cdot\int_{T_k^q}\rho'^q_k\frac{1}{R}\mathrm{d}S'\mathrm{d}S\\ &= |\mathbf{r}_1-\mathbf{r}_3|^2 I_{11}+|\mathbf{r}_2-\mathbf{r}_3|^2 I_{22}+2(\mathbf{r}_1-\mathbf{r}_3)\cdot(\mathbf{r}_2-\mathbf{r}_3)I_{12}+(\mathbf{r}_3-\mathbf{r}_m)\cdot(\mathbf{r}_3-\mathbf{r}_k)\\ &+ (\mathbf{r}_1-\mathbf{r}_3)\cdot(2\mathbf{r}_3-\mathbf{r}_m-\mathbf{r}_k)I_1+(\mathbf{r}_2-\mathbf{r}_3)\cdot(2\mathbf{r}_3-\mathbf{r}_m-\mathbf{r}_k)I_2 \qquad (3.138)\end{aligned}$$

where the exact values of I_0, $I_{1(2)}$, $I_{11(22)}$, and I_{12} have been given, respectively, by (5), (6), (7), and (8) in [88] as functions of the triangle sidelengths.

3.7 Algebraic Stability Analysis

Considering EM wave scattering from a perfect conducting body, the governing TDIE is projected through taking the inner product onto a finite dimensional linear vector space,

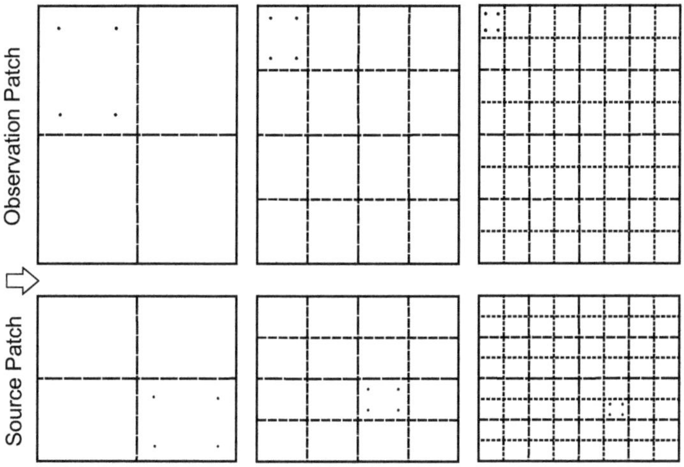

Figure 3.14: A typical n-fold concurrently partitioned quadrilateral subdomains, $n = 3$.

spanned by N_s spatial basis functions, and solved by the MOT, MOD, or FDDM schemes for up to $n = N_t$ time (order) steps, i.e.,

$$\hat{\mathbf{n}} \times \left(\hat{\mathbf{n}} \times [\mathbf{E}^i(\mathbf{r},t) + \mathbf{E}^s(\mathbf{r},t)]\right) = 0 \quad \mapsto \quad \bar{\bar{Z}}_0 \bar{I}_n = \bar{V}_n - \sum_{r=b}^{n-1} \bar{\bar{Z}}_{n-r} \bar{I}_r - \bar{\gamma}_{n-1} \tag{3.139}$$

where $b = 1$, $\bar{\gamma}_{n-1} = 0$, and $\bar{\bar{Z}}_n = 0$ after $n > N_g$ for the MOT whereas $b = 0$ for the MOD and FDDM cases. The matrix equation (3.139) can be reordered as

$$\begin{bmatrix} \bar{\bar{Z}}_0 & \bar{\bar{O}} & & & \\ \bar{\bar{O}} & \bar{\bar{I}} & \bar{\bar{O}} & & \\ \bar{\bar{O}} & \bar{\bar{O}} & \bar{\bar{I}} & \bar{\bar{O}} & \\ \vdots & & & \ddots & \\ \bar{\bar{O}} & \cdots & & \bar{\bar{O}} & \bar{\bar{I}} \end{bmatrix} \begin{bmatrix} \bar{I}_n \\ \bar{I}_{n-1} \\ \bar{I}_{n-2} \\ \vdots \\ \bar{I}_1 \end{bmatrix} = \begin{bmatrix} \bar{\bar{Z}}_1 & \bar{\bar{Z}}_2 & \bar{\bar{Z}}_3 & \cdots & \bar{\bar{Z}}_n \\ \bar{\bar{I}} & \bar{\bar{O}} & & & \\ \bar{\bar{O}} & \bar{\bar{I}} & \bar{\bar{O}} & & \\ \vdots & & & \ddots & \\ \bar{\bar{O}} & \cdots & \bar{\bar{O}} & \bar{\bar{I}} & \bar{\bar{O}} \end{bmatrix} \begin{bmatrix} \bar{I}_{n-1} \\ \bar{I}_{n-2} \\ \bar{I}_{n-3} \\ \vdots \\ \bar{I}_0 \end{bmatrix} \tag{3.140}$$

or equivalently a system of linear equations $\boxed{\bar{\mathbf{I}}^n = \bar{\bar{\mathbf{Y}}}\,\bar{\mathbf{I}}^{n-1}}$ with size $N_t N_s$ and the normalized top row block including $\bar{\bar{Y}}_n = \bar{\bar{Z}}_0^{-1} \bar{\bar{Z}}_n$.

$$\begin{bmatrix} \bar{I}_n \\ \bar{I}_{n-1} \\ \bar{I}_{n-2} \\ \vdots \\ \bar{I}_1 \end{bmatrix} = \begin{bmatrix} \bar{\bar{Y}}_1 & \bar{\bar{Y}}_2 & \bar{\bar{Y}}_3 & \cdots & \bar{\bar{Y}}_n \\ \bar{\bar{I}} & \bar{\bar{O}} & & & \\ \bar{\bar{O}} & \bar{\bar{I}} & \bar{\bar{O}} & & \\ \vdots & & & \ddots & \\ \bar{\bar{O}} & \cdots & \bar{\bar{O}} & \bar{\bar{I}} & \bar{\bar{O}} \end{bmatrix} \begin{bmatrix} \bar{I}_{n-1} \\ \bar{I}_{n-2} \\ \bar{I}_{n-3} \\ \vdots \\ \bar{I}_0 \end{bmatrix} \tag{3.141}$$

In the MOT scheme, the tail of the retarded current and thereby the system matrix in (3.140) extend respectively up to $n - N_g$ and N_g of $\mathcal{O}(\frac{R_{\max}}{c\Delta t})$.

$$\begin{bmatrix} \bar{\mathrm{I}}_n \\ \bar{\mathrm{I}}_{n-1} \\ \bar{\mathrm{I}}_{n-2} \\ \vdots \\ \bar{\mathrm{I}}_{1+n-N_g} \end{bmatrix} = \begin{bmatrix} \bar{\bar{\mathrm{Y}}}_1 & \bar{\bar{\mathrm{Y}}}_2 & \bar{\bar{\mathrm{Y}}}_3 & \cdots & \bar{\bar{\mathrm{Y}}}_{N_g} \\ \bar{\bar{\mathrm{I}}} & \bar{\bar{\mathrm{O}}} & & & \\ \bar{\bar{\mathrm{O}}} & \bar{\bar{\mathrm{I}}} & \bar{\bar{\mathrm{O}}} & & \\ \vdots & & & \ddots & \\ \bar{\bar{\mathrm{O}}} & \cdots & \bar{\bar{\mathrm{O}}} & \bar{\bar{\mathrm{I}}} & \bar{\bar{\mathrm{O}}} \end{bmatrix} \begin{bmatrix} \bar{\mathrm{I}}_{n-1} \\ \bar{\mathrm{I}}_{n-2} \\ \bar{\mathrm{I}}_{n-3} \\ \vdots \\ \bar{\mathrm{I}}_{n-N_g} \end{bmatrix} \tag{3.142}$$

The eigenvalue spectrum of the companion matrix $\bar{\bar{\mathbf{Y}}}$ determines the stability behaviour of the TDIE. In the stability analysis of the PDE $\frac{\partial}{\partial t}\mathbf{J} = \bar{\bar{\mathbf{Z}}}\cdot\mathbf{J} + \bar{\mathrm{V}}(t)$ with the system matrix $\bar{\bar{\mathbf{Z}}}$, similarly the matrix relating the present response of the system $\mathbf{J}_n$ with its previous status $\mathbf{J}_{n-1}$ depends on the time integration scheme and is called the iteration (amplification) matrix $\bar{\bar{\mathbf{Y}}}$. The eigenvalues λ of the system and iteration matrices are related through $\lambda_{\mathbf{Y}} = (1 + \Delta t \lambda_{\mathbf{Z}})$. The eigenvalue distribution of iteration matrix in the FIT or the FDTD methods for solving the ODE by different time integration schemes demonstrate that the popular Crank-Nicolson and leap-frog algorithms are a symplectic time integrator as the associated eigenvalues of its iteration matrix lie on the unit circle, i.e. $|\lambda| = 1$. Similar constellations are obtained for finite volume method (FVM) with first order central flux [12].

To approximate the time derivative in TDIE, the frequently applied stable implicit backward Euler time integration method is inherently dissipative and it can not preserve the EM energy. The second order Crank-Nicolson method is an energy conserving time integration scheme for the most PDE, as it approximates the time derivative by the central difference and applies time averaging on the retarded integrals. However, as the distribution of eigenvalues in the complex plane shows in Section 6.10, the conservation property of the Crank-Nicolson integrator when applied in the classical MOT fails owing to the numerical approximations.

Chapter 4

Accelerated Solvers

4.1 Comparison of MOT and MOD Methods

Commonly, the MOT schemes are the primary candidate to solve the TDIEs numerically [90]. As discussed in 6.2, the MOT recipes, however, suffer from late-time instability. Much work has been done to postpone or filter out the occurrence of exponentially growing fluctuations on the tail of response. Nonetheless, among robust generations of TDIE-based solvers such as the FDDM, the MOD schemes [91] are solely the only TDIE methods that are always stable [10]. The MOT algorithms use locally supported (commonly shifted Lagrange) expansion functions and point matching testing in time with interpolation in between past nodal points, Section 3.1. In the MOD methods, as explained in Section 3.4, a set of orthogonal entire-domain basis functions, namely the weighted Laguerre polynomials, together with the Galerkin's method for temporal testing are used. Thus, the MOD represents the temporal variation of the unknowns more smoothly than MOT does, except for few initial orders of Laguerre polynomials that contribute dominantly to the very early stages. As a result, the MOT methods are not as accurate as the MOD methods [92]. Although both handle the time integrations (3.50) and derivatives (3.48), (3.49) fully analytically, the orthogonality in MOD methods let the time variable to be integrated out (3.66)-(3.68). In contrary to the non-energy conserving MOT methods, the complete set of the Laguerre expansion resembles a symplectic time integration method, when sufficiently large number of time bases N are marched, Section 3.6. This also implies that the MOD methods provide better accuracy [73] whereas their computational burdens outweigh the MOT. In the advanced version of the MOD (AMOD), introduced in Section 3.4.3, the temporal testing is performed before the spatial testing whereby the unrealistic assumption of no changes for the unknown transient quantity within the subdomains is avoided and the accuracy is improved.

The dominant cost of the MOT, FDDM, and (A)MOD methods are mainly the computation of past solution couplings involving space-time convolution of the induced currents with the Green's function. The computational cost of the classical MOT schemes scales as $\mathcal{O}(N_g N_t N_s^2)$, where N_t and N_s (equivalently N and M in Chapter 3) denote the number of subdomain temporal and spatial basis functions, respectively, and N_g is the maximum number of the last retarded time steps in which the scatterer subdomains interact (3.20). Depending on the sizes of the scatterer, mesh, and time step as well as frequency content of the excitation, the longest tail of the delayed samples N_g may vary from $N_g << N_t$ to N_g of $\mathcal{O}(N_t)$, e.g., for planar surfaces N_g is typically of $\mathcal{O}(\sqrt{N_s})$. In the FDDM and MOD methods, independent from the problem at hand always $N_g \geq N_t$ is obtained which apparently rises the total computational cost to $\mathcal{O}(N_t^2 N_s^2)$. In other words, solving nondispersive unbounded media by the MOD methods due to the development of infinite temporal tail has the same computational cost as the MOT schemes in the extreme case of a very strong dispersion or in lossy media [93]. Generally, large time step sizes improve the late-time stability of the MOT, which implies smaller N_g. On the other hand, the MOD may call for accumulation of plenty of orders N_t to diminish early ripples in the initial coarse representations of the system response [94]. However, N_t in the MOT and MOD methods are not conceptually the same. In the MOD methods, the temporal degrees of freedom of the surface current density N_t represents the time-bandwidth product of the waveforms to be approximated, see App. 8.5.1. Moreover, in the MOD, $\mathcal{O}(N_t)$ inner loop operations are applied on the past current vectors to calculate the integration or differentiation operations. It is reminded that more than six new marching-without-time-variable recipes have been introduced in Section 3.4.4 to omit these existing $\mathcal{O}(N_t)$ vector summations on the behind polynomial orders. Thus, the cost of the MOD methods with reduced sums is larger than the MOT when the number of time steps is not too large.

The MOT solvers have been accelerated by the plane wave time domain (analogous to the fast multiple method in frequency domain) [27] and fast Fourier transform (FFT)-based algorithms [90]. The generalization of the former to dispersive and/or layered media is nontrivial. In the FFT-based TDIE methods, the convolution products are calculated based on (Toeplitz)-block-Toeplitz properties (constant along diagonals) of the impedance matrices [11, 65]. In [95], the unknowns of uniformly meshed planar structures are separated into the local x and y components (i.e., four mutual orientations) and a two-dimensional (2-D) FFT has been exploited for circular convolutions. On the other hand, the periodicity-based moment method (MoM) takes advantages of similar properties to analyze large finite antenna arrays [96]. Nevertheless, the archival literature suffers from the lack of detailed guidelines on amalgamation of the Toeplitz arrangements due to the system periodicity and uniform meshing. The gap is well filled in Section 4.3 by the multi-level Toeplitz matrix-vector multiply algorithm in [97].

Since the physics underlying the scattering process is time invariant, when the time variable in the TDIE is discretized uniformly, the retarded matrices form a Toeplitz system in the temporal dimension. The MOT solvers have been also accelerated by the time-FFT-based algorithms [93] in which the convolution products are calculated based on Toeplitz properties of the (aggregates of) impedance matrices. The time-FFT proposed in [65] is a global solution scheme, i.e., it does not march in time (as the method of lines does) but instead it solves for all space-time unknowns simultaneously (as the Rothe's method does), whereas the FFT-accelerated marching solution methods provide an appealing avenue for tackling long time simulation of large scale scattering problems. The explanation of the time-domain adaptive integral method (AIM) [98, 90] connotes using what the authors call blocked four-dimensional (4-D) FFTs to combine the implicit temporal arrangement by the 3-D forward spatial FFTs. This implies that the uniform AIM grid has to be zero padded at least to the double size of the auxiliary sources in each dimension to avoid aliasing errors. This chapter guides to incorporate the space-time convolution products of the TDIE methods in single 1-D FFT.

To summarize briefly, on each recursion level of the solution procedure, the computational work is dominated by the matrix-vector multiplications not by matrix equation solution and the block-Toeplitz property of the individual and group of matrices can be interpreted as space and time convolutions respectively to be emerged from space/time-FFT.

The FFT process, however, perishes the sparsity of the MOT matrices. The hierarchical grouping of sparse interactions has been suggested to alleviate the redundancy in FFT convolution of sparse matrices in the MOT methods [29]. On the other hand, the MOD methods generate dense matrices and, as explained above, they appeal more than the MOT methods for exploiting auxiliary techniques to reduce the CPU cycles and memory demands. Nevertheless, no accelerating or compression technique has yet been introduced for the MOD recipes. Fortunately, interaction matrices in the (A)MOD methods can also be arranged in a two-level block-Toeplitz form due to the translationally invariant nature of the Green's function. Based on this property, the next three sections introduce an efficient FFT algorithm for matrix compression and fast matrix-vector multiplication in solving the surface integral equations pertinent to the analysis of (periodic) planar or rotationally symmetric structures by the AMOD. The Toeplitz properties due to the space periodicity and uniform meshing are merged together in a multi-level fashion in Section 4.3. The algorithms in Section 4.3 and Section 4.2 reduce the serial complexities and storage requirements respectively to $\mathcal{O}(N_t^2 N_s \log N_s)$ and $\mathcal{O}(N_t N_s)$. It is proven that without taking into account any special spatial properties of the matrix blocks, the MOD solution utilizing the Toeplitz property in time in Section 4.4 also achieves to the logarithmic complexity scaling $\mathcal{O}(N_t \log^2 N_t)$. Recently, the subdivision of large-size Toeplitz block aggregates to elementary matrix blocks was proposed to speed up the MOT scheme [99]. The present work investigates the performance of the alternative techniques on the MOT, MOD, and FDDM methods. It is shown that for large N_s, the subdivision to fixed-size blocks alone becomes inferior to the native multilevel aggregate matrix-vector multiply when the MOD in conjunction with mapping to the uniform grid is used. Finally, a hybrid grouping approach is formulated in which the complexity and memory requirements are further reduced by avoiding repeating the FFT computation for current vectors. Principal aspects of implementation are discussed in this chapter as well.

4.2 Space Convolution Products

The tangential component of the scattered fields constitute the left-hand sides of the TDIEs, namely in (2.10) or (2.11), and (2.24). Principally, $\mathbf{J}(\mathbf{r}, t)$ is convolved by the Green's function to generate the scattered fields. The free space Green's function $\frac{\delta(t-R/c)}{R}$ and accordingly its time convolution with the weighted temporal expansion functions $\frac{I_\nu(s\frac{R}{c})}{R}$ are translational invariant, i.e., they are a function of $R_{mk} = |\mathbf{r}_m - \mathbf{r}'_k|$, the distance between the observation and source points. Therefore, when S is uniformly meshed, the dense and possibly asymmetric square matrices $\{\bar{\bar{Z}}_{i-j}\}_{m,k}$ in (3.72)-(3.74) can be represented only by its unique entries $\{\bar{Z}_\nu\}_{m-k}$. The same holds true for $\{\bar{\bar{Z}}_{n-r}\}_{m,k}$ in (3.21) or (3.22). In other words, the matrices $\bar{\bar{Z}}_\nu$ are (Toeplitz)-block-Toeplitz, and hence, the matrix-vector products on the right-hand side of (3.19) are convolution products and they can be efficiently calculated via element-by-element multiplication in the spectral domain as

$$\sum_{j=0}^{i-1} \bar{\bar{Z}}_\nu \bar{I}_j = \sum_{j=0}^{i-1} \ddagger \left(\bar{Z}_\nu \otimes \underline{\bar{I}}_j\right) = \dagger \mathrm{Re}\{\mathrm{FFT}^{-1} \left(\sum_{j=0}^{i-1} \mathrm{FFT}\{\bar{Z}_\nu\} \cdot \mathrm{FFT}\{\underline{\bar{I}}_j\} \right) \} \tag{4.1}$$

where $\bar{Z}_\nu$ is a vector consisting unique entries of blocks in $\bar{\bar{Z}}_\nu$ and the auxiliary vector $\underline{\bar{I}}_j$ is the flipped up/down and zero-padded extension of $\bar{I}_j$ with the same size as $\bar{Z}_\nu$. Operators

‡ and † extract only the desired entries of the product, namely †Re{} flips the resulting sequence in down/up direction and picks up the real part of those array elements located corresponding to the positions of the original nonzero entries in $\underline{\mathrm{I}}_j$. The same procedure is applied to the matrix-vector products of $\bar{\bar{\mathrm{Z}}}_\nu^{\mathrm{A}}\bar{\mathrm{I}}_\iota$ and $\bar{\bar{\mathrm{Z}}}_\nu^{\phi}\bar{\mathrm{I}}_\iota$ in (3.72) where

$$\left[\mathrm{Z}_\nu^{\mathrm{A}}\right]_{mk} = s\frac{\mu}{4\pi}\int_S \mathbf{f}_m(\mathbf{r})\cdot\int_S \frac{I_\nu(s\frac{R}{c})}{R}\mathbf{f}_k(\mathbf{r}')\mathrm{d}S'\mathrm{d}S \tag{4.2}$$

$$\left[\mathrm{Z}_\nu^{\phi}\right]_{mk} = \frac{4}{s}\frac{1}{4\pi\epsilon}\int_S \nabla_{\mathbf{r}}\cdot\mathbf{f}_m(\mathbf{r})\int_S \frac{I_\nu(s\frac{R}{c})}{R}\nabla_{\mathbf{r}'}\cdot\mathbf{f}_k(\mathbf{r}')\,\mathrm{d}S'\mathrm{d}S \tag{4.3}$$

or $\dot{\bar{\bar{\mathrm{Z}}}}_\nu\bar{\mathrm{I}}_j$ and $\bar{\bar{\mathrm{Z}}}_\nu^{\Phi}\bar{\mathrm{I}}_\iota$ in (3.73) where $\bar{\bar{\mathrm{Z}}}_\nu^{\Phi} = s\,\bar{\bar{\mathrm{Z}}}_\nu^{\phi}$ or $\bar{\bar{\mathrm{Z}}}_\nu^{A}\bar{\mathrm{I}}_\iota$ in (3.74). The computational expenses of evaluating the double surface quadratures in Section 3.6.1 are relatively high. Hence, the compressed versions of $\bar{\bar{\mathrm{Z}}}_\nu$, $\bar{\mathrm{Z}}_\nu$ with dimensions of $\mathcal{O}(N_s)$, are stored in memory for further usage in constructing (3.72)-(3.74) or (3.22) through (4.1) and solving it for next higher orders of i or time step n. Besides, when symmetrical quadrature routines are used to numerically calculate the double surface integrals over non-overlapped source and observation subdomains in Section 3.6.1, $\{\bar{\mathrm{Z}}_\nu\}_{m-k}$ reduce to $\{\bar{\mathrm{Z}}_\nu\}_{|m-k|}$, i.e., the number of unique entries N_u halves and the cost of products may be further reduced.

To better explain fast computation of the spatial convolution products, we consider an inclined wire antenna modeled by a narrow strip, on which the current distribution has been approximated by N_s+1 rectangular surface patches. Approximating the outer integrals, e.g. in (4.2) and (4.3), by the value of the integrands at the center of observation patch, $\bar{\bar{\mathrm{Z}}}_\nu\bar{\mathrm{I}}_\mathrm{j}$ find the following pattern

$$\begin{bmatrix} Z_0 & Z_1 & \cdots & Z_{N_s-1} \\ Z_{-1} & Z_0 & \cdots & Z_{N_s-2} \\ \vdots & \vdots & \ddots & \vdots \\ Z_{1-N_s} & Z_{2-N_s} & \cdots & Z_0 \end{bmatrix}\begin{bmatrix} I_1 \\ I_2 \\ \vdots \\ I_{N_s}\end{bmatrix}. \tag{4.4}$$

Thus,

$$\begin{aligned} \bar{\mathrm{Z}}_\nu &= [Z_{1-N_s}\ Z_{2-N_s}\ \cdots\ Z_{N_s-2}\ Z_{N_s-1}]_{1\times N_u} \\ \underline{\mathrm{I}}_j &= [I_{N_s}\ I_{N_s-1}\ \cdots\ I_2\ I_1\ 0\ 0\ \cdots\ 0]_{1\times N_u} \end{aligned}$$

where $N_u = 2N_s - 1$. As the second example, assume a tapered transmission line consisting unparallel microstrips each paved with N_s+1 rectangle subdomains defining N_s rooftop basis functions, the associated portion of the impedance matrices form two Toeplitz constellations as follows:

$$\bar{\bar{\mathrm{Z}}}_\nu\bar{\mathrm{I}}_\mathrm{j} = \begin{bmatrix} \mathbf{Z} & \mathbf{Z}' \\ \mathbf{Z}' & \mathbf{Z}\end{bmatrix}\begin{bmatrix}\mathbf{I} \\ \mathbf{I}'\end{bmatrix} \tag{4.5}$$

where the submatrices $\mathbf{Z}^{(')}$ and subarrays $\mathbf{I}^{(')}$ respectively have the same structure as the asymmetric matrix and vector on (4.4). As a result,

$$\begin{aligned} \bar{\mathrm{Z}}_\nu &= \left[Z_{1-N_s}\ \cdots\ Z_{N_s-1}\ Z'_{1-N_s}\ \cdots\ Z'_{N_s-1}\right] \\ \underline{\mathrm{I}}_j &= \left[I'_{N_s}\ \cdots\ I'_1\ 0\ \cdots\ 0\ I_{N_s}\ \cdots\ I_1\ 0\ \cdots\ 0\right]. \end{aligned}$$

The number of intermediately inserted zeros is $N_s - 1$, one less than the separation length of the Toeplitz blocks, e.g. when $N_s = 4$

$$\bar{\bar{\mathrm{Z}}}_\nu \bar{\mathrm{I}}_\mathrm{j} = \begin{bmatrix} Z_0 & Z_1 & Z_2 & Z_3 & Z_4 & Z_5 & Z_6 & Z_7 \\ Z_{-1} & Z_0 & Z_1 & Z_2 & Z_{-5} & Z_4 & Z_5 & Z_6 \\ Z_{-2} & Z_{-1} & Z_0 & Z_1 & Z_{-6} & Z_{-5} & Z_4 & Z_5 \\ Z_{-3} & Z_{-2} & Z_{-1} & Z_0 & Z_{-7} & Z_{-6} & Z_{-5} & Z_4 \\ Z_4 & Z_5 & Z_6 & Z_7 & Z_0 & Z_1 & Z_2 & Z_3 \\ Z_{-5} & Z_4 & Z_5 & Z_6 & Z_{-1} & Z_0 & Z_1 & Z_2 \\ Z_{-6} & Z_{-5} & Z_4 & Z_5 & Z_{-2} & Z_{-1} & Z_0 & Z_1 \\ Z_{-7} & Z_{-6} & Z_{-5} & Z_4 & Z_{-3} & Z_{-2} & Z_{-1} & Z_0 \end{bmatrix} \begin{bmatrix} I_1 \\ I_2 \\ I_3 \\ I_4 \\ I_5 \\ I_6 \\ I_7 \\ I_8 \end{bmatrix}$$

$$\begin{aligned} \bar{\mathrm{Z}}_\nu &= [Z_{-3}\, Z_{-2}\, Z_{-1}\, Z_0\;\; Z_1\, Z_2\, Z_3\, Z_{-7}\, Z_{-6}\, Z_{-5}\, Z_4\, Z_5\, Z_6\, Z_7] \\ \underline{\bar{\mathrm{I}}}_j &= [I_8\;\; I_7\;\; I_6\;\; I_5\;\; 0\;\; 0\;\; 0\;\; I_4\;\; I_3\;\; I_2\;\; I_1\;\; 0\;\; 0\;\; 0]\,. \end{aligned}$$

For a parallelogram sheet partitioned by $(N_x + 1) \times (N_y + 1)$ series of parallelogram patches whose corresponding edges have been numbered sequentially, the impedance matrices are (can be ordered in the form of) such four Toeplitz-block-Toeplitz submatrices,

$$\bar{\bar{\mathrm{Z}}}_\nu \bar{\mathrm{I}}_\mathrm{j} = \begin{bmatrix} \hat{\mathbf{Z}}_{P\times P} & \grave{\mathbf{Z}}_{P\times Q} & \\ \acute{\mathbf{Z}}_{Q\times P} & \check{\mathbf{Z}}_{Q\times Q} & \vdots \\ & \cdots & \end{bmatrix} \begin{bmatrix} \hat{\mathbf{I}}_{P\times 1} \\ \check{\mathbf{I}}_{Q\times 1} \\ \vdots \end{bmatrix} \tag{4.6}$$

where $P = N_x(N_y + 1)$, $Q = N_x N_y$, and the 2-level block-Toeplitz submatrices $\hat{\mathbf{Z}}$, $\grave{\mathbf{Z}}$, $\acute{\mathbf{Z}}$, and $\check{\mathbf{Z}}$ contain repeated blocks of size $N_x \times N_x$, each with pattern similar to that of (4.4), Fig. 4.1. Therefore, the product of the four submatrices by the corresponding two subarrays can be obtained through (4.1) including 4 parallel FFT executions with respective lengths of $\hat{N}_u = (2N_x - 1)(2N_y + 1)$, $\acute{N}_u = \grave{N}_u = (2N_x - 1)(2N_y)$, and $\check{N}_u = (2N_x - 1)(2N_y - 1)$. The remaining N_y rows and columns as well as the rest in the lower left corner of $\bar{\bar{\mathrm{Z}}}_\nu$ relating interactions with possibly non-uniformly meshed parts of the body are multiplied in the conventional way. Here, the rooftop edges are not indexed by canonical numbering along the two distinct directions, but rather to gain the block-Toeplitz characteristic for $\acute{Z}$ and $\grave{Z}$ additionally, the numbering of unparallel groups of edges is counted one group after the other. Identification numbers are assigned first to the codirectional edges oriented in the larger dimension, that is the dimension with more divisions so to say $N_x \geq N_y$, excluding N_y horizontal ending edges that are enlisted at the end. Thus, there is no need to transfer the plate geometry to the xy-plane and anchor its corner to the origin and canonically numbering in y direction as suggested by [95]. Of course, $\acute{\mathbf{Z}}_{Q\times P}$ and $\grave{\mathbf{Z}}_{P\times Q}$ parts of $\bar{\bar{\mathrm{Z}}}_\nu^A$ and $\bar{\bar{\mathrm{Z}}}_\nu^\Phi$ are zero due to the orthogonal orientation of spatial bases in (2.46).

Uniform discretization of cylindrical parts of the scatterer for proper local alignment of the surface normal vectors $\hat{\mathbf{n}}$, e.g. modeling tube-like parts of the structure such as accelerator beam tubes, via interconnects, etc. by the rooftop bases has been proposed in Section 2.4.3. The sequential edge indexing of (inclined) cylinders parts with rotationally (a)symmetric cross sections directly renders interaction matrices containing Toeplitz-block-Toeplitz shaped submatrice(s) associated to the mutual coupling of the rooftop bases on the tube parts. Considering a tube is partitioned into N_ϕ subdomains in azimuthal and N_z subdomains in longitudinal directions, $P = N_\phi N_z$ and $Q = N_\phi(N_z - 1)$ in (4.6) and

$\hat{N}_u = (2N_\phi - 1)(2N_z - 1)$, $\acute{N}_u = \grave{N}_u = (2N_\phi - 1)(2N_z - 2)$, and $\check{N}_u = (2N_\phi - 1)(2N_z - 3)$. Here also $\acute{\mathbf{Z}}^A$ and $\grave{\mathbf{Z}}^A$ are zero. Fig. 4.1 typically illustrates how the compression algorithm serially puts in order the unique entries of block aggregates to fill-in $\bar{\mathrm{Z}}_\nu$ for a resulting 2-level block-Toeplitz submatrix.

18	**19**	**20**	**21**	**25**	**26**	**27**	**28**	**32**	**33**	**34**	**35**	**39**	**40**	**41**	**42**
17	18	19	20	**24**	25	26	27	**31**	32	33	34	**38**	39	40	41
16	17	18	19	**23**	24	25	26	**30**	31	32	33	**37**	38	39	40
15	16	17	18	**22**	23	24	25	**29**	30	31	32	**36**	37	38	39
11	**12**	**13**	**14**	18	19	20	21	25	26	27	28	32	33	34	35
10	11	12	13	17	18	19	20	24	25	26	27	31	32	33	34
9	10	11	12	16	17	18	19	23	24	25	26	30	31	32	33
8	9	10	11	15	16	17	18	22	23	24	25	29	23	31	32
4	**5**	**6**	**7**	11	12	13	14	18	19	20	21	25	26	27	28
3	4	5	6	10	11	12	13	17	18	19	20	24	25	26	27
2	3	4	5	9	10	11	12	16	17	18	19	23	24	25	26
1	2	3	4	8	9	10	11	15	16	17	18	22	23	24	25

Figure 4.1: Positions of the unique elements of the submatrix $\acute{\mathbf{Z}}$ ($P = 16, Q = 12$) that can be generated by a parallelogram plate ($N_x = 4$, $N_y = 3$) or inclined cylinder ($N_\phi = 4$, $N_z = 5$) case studies. The 2nd level of Toeplitz property has been highlighted by the dashed lines encompassing the blocks. The numbering of elements infers the calling sequence in constructing $\bar{\mathrm{Z}}_\nu$. Periodically $N_x - 1 = 3$ zeros are inserted between every N_x consecutive current elements (level borders) to build $\underline{\bar{\mathrm{I}}}_j$.

In general case, according to the block-Toeplitz structure of the submatrices, zeros are first inserted at appropriate locations [97] into the reversed version of $\bar{\mathrm{I}}_j$ so as to obtain the auxiliary current vector with proper alignment, suitable for direct convolution with $\bar{\mathrm{Z}}_\nu$. The auxiliary vector $\underline{\bar{\mathrm{I}}}_j$ is then zero padded to the length of $\bar{\mathrm{Z}}_\nu$ before the FFT and subsequent multiplication in Fourier domain. The convolution is readily accomplished in $\mathcal{O}(N_s \log N_s)$ operations. The location of initially inserted zeros are then used directly to suppress the extra terms and recovering the reconstructed product in the final step. Note that some algorithms, such as the one proposed by [11], only exploit the Toeplitz structure of blocks, rather the block-Toeplitz property in companion as explained here. In addition, it is shown in Fig. 4.1 that the present extended algorithm inspired from [97] can also be applied to rectangular matrices $\acute{\mathbf{Z}}$ and $\grave{\mathbf{Z}}$ when $P \neq Q$.

4.3 Periodicity and Multilevel Toeplitz Matrices

Let S consist of duplications of a cell S_0 at regularly space positions with $\mathbf{D}_p$ centric spacing where the integer p is the subsystem repetition labels in alignment with the specific direction $\hat{\mathbf{D}}$. The interaction matrix elements can then be computed from (4.3) and (4.2) alternatively in local groups (p, q) by considering

$$\int_S (\nabla_{\mathbf{r}}\cdot)\mathbf{f}_m(\mathbf{r}) \cdot \int_S \frac{I_\nu(s\frac{R}{c})}{R}(\nabla_{\mathbf{r}'}\cdot)\mathbf{f}_k(\mathbf{r}')\mathrm{d}S'\mathrm{d}S =$$
$$\int_{S_0} (\nabla_{\mathbf{r}}\cdot)\mathbf{f}_m(\mathbf{r}) \cdot \int_{S_0} \frac{I_\nu(s\frac{R_d}{c})}{R_d}(\nabla_{\mathbf{r}'}\cdot)\mathbf{f}_k(\mathbf{r}')\mathrm{d}S'\mathrm{d}S, \qquad (4.7)$$

where $R_d = |\mathbf{r} - \mathbf{r}' + \mathbf{D}_{p-q}|$ in which the variation range of the global coordinates $\mathbf{r}$ and $\mathbf{r}'$ are confined to the primary subsystem S_0. The dependency on $\mathbf{D}_{p-q}$ substantiates an (additive) Toeplitz property in $\bar{\bar{\mathbf{Z}}}_\nu$ when all identically ordered unknowns corresponding to the cells along $\hat{\mathbf{D}}$ are listed sequentially one after the other. Fig. 4.2 exhibits the fractal-like pattern of the interaction matrices $\bar{\bar{\mathbf{Z}}}_\nu$ for a set of periodic non-uniformly meshed objects with 4-level Toeplitz property on the fundamental blocks independent from the island meshing, two interior levels due to the inherent subsystem periodicity along the locally assumed horizontal $\hat{\mathbf{x}}$ and vertical $\hat{\mathbf{y}}$ axes ($n_x = 2$, $n_y = 2$) and two additional outer levels owing to $n_{yy} = 3$ and $n_{xx} \geq 4$ times $n_x n_y$-cell group replication along the $\hat{\mathbf{y}}$ and $\hat{\mathbf{x}}$ directions, respectively. Therefore, the multiple geometry repetition levels can be transferred to multiple repetition levels in the matrix structure. Matrix-vector multiply can be computed using the FFT approach (4.1) in block-wise form, that is corresponding elements from every block are multiplied collectively, so as to scale the complexity to $\mathcal{O}(N_t^2 N_s^2 \frac{1}{n_{xy}} \log n_{xy})$ where $n_{xy} = n_x n_y n_{xx} n_{yy}$. The outermost corner subblocks (ending branches) in Fig. 4.2 are related to the far subsystems, and they may be approximated equally by the interaction of centric elements. It is worth mentioning that the present algorithm does not demand any edge and corner sub-entire domain basis functions or dummy cells as in [100].

Assuming that a 2-D periodic rectangular-shape PEC patches (a capacitive mesh filter) with finite size of $n_x n_y$ cells is meshed by $n_x n_y N_{s0}$ rooftop basis functions (where $N_{s0} = (N_x + 1)N_y + N_x(N_y + 1)$), the submatrices in (4.6) are expanded to $P = n_x n_y\, N_x(N_y + 1)$, $Q = n_x n_y\, N_x N_y$. Resembling the former nested periodic case in Fig. 4.2, the encompassed block-wise Toeplitz property because of the periodicity can be concatenated to the every Toeplitz-block-Toeplitz interaction submatrices of the underlying uniformly meshed subsystemes in (4.6) when the periodicity effects are exerted in the most outer Toeplitz levels. Fig. 4.3 illustrates how the periodicity along two different axes can be incorporated within the 3rd and 4th levels of the four block-Toeplitz submatrices when the subsystems are composed of any either of the generic case studies in Fig. 4.1. Dots in Fig. 4.3 specify the primal location of the required elements to be calculated and emplaced in $\bar{\mathrm{Z}}_\nu$ for the first quarter of the matrices. Starting from the lower-left corner entry in every one of the four submatrices, among the unique elements, those whose column and row indices are respectively greater and smaller than the others are emplaced prior to the rest in the auxiliary array sequence. This is equivalent to include only the elemental coupling of the basis functions while moving first toward the positive $\hat{\mathbf{x}}$ axis and then along $\hat{\mathbf{y}}$ in each cell and considering the mutual couplings between the primal cell and the repeated copies afterwards. To fill

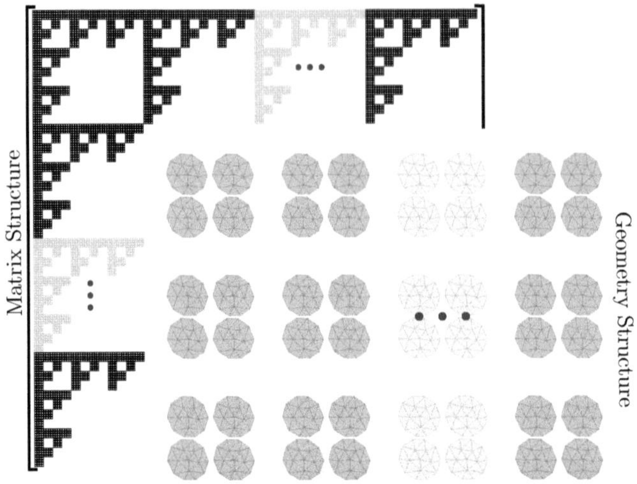

Figure 4.2: All (partially) filled square blocks are lined up corresponding to the distances between the (groups of) array elements. The matrix is fully dense and only the replicas of the original subblocks have not been depicted to reflect the shift invariancy (diagonal displacement) of the (sub)blocks on the four outermost nested Toeplitz levels.

the auxiliary current vector $\underline{\bar{I}}_j$, in transition to one higher level, different number of zeros has to be inserted between the corresponding current elements [97]. The three different stages happen between every N_x, n_yN_x, $n_xn_yN_x$ (that is every 4, 12, 36) elements of the flipped current vector, where respectively $N_x - 1$, $\lfloor \frac{\hat{N}_u}{2} \rfloor$, $n_y\hat{N}_u - \lfloor \frac{\hat{N}_u}{2} \rfloor$ (in the example 3, 17, 87) zeros have to be inserted in between. The obtained vector $\underline{\bar{I}}_j$ are finally zero-padded up to the length $n_xn_y\hat{N}_u$. Here also $\lfloor x \rfloor$ denotes the greatest integer less than or equal to x. The first submatrix-vector product can be retrieved following the inverse FFT, once after every $N_x - 1$, $\lfloor \frac{\hat{N}_u}{2} \rfloor$, $n_y\hat{N}_u - \lfloor \frac{\hat{N}_u}{2} \rfloor$ elements of the flipped array, correspondingly N_x, n_yN_x, $n_xn_yN_x$ redundant elements are skipped.

Although for the ease of explanation the ordered selection of the unique matrix elements is explicated, no element picking-up procedure is to be developed in practice, rather the unique interactions of the basis functions are directly addressed for proper alignment in the auxiliary vectors and there is no need to build the complete (sub)matrices. In the phased array antennas, frequency-selective surfaces (FSS), photonic bandgap materials, artificial left-handed materials (metamaterials), etc. where the unit cell is composed of arbitrarily shaped patches, the auxiliary uniform meshes of the AIM in Section 4.6 can be interrelated with the present algorithm. Here, the advantage is put on the finite structures like patch antennas. Infinite repetitions are preferably solved by the unit-cell approach [101]. A similar principle can be applied to a 3-D uniform auxiliary grid [98] that encases every scatterer cell as well as 3-D periodicity (an extra outer level with $2n_z - 1$ blocks) which culminates the nested Toeplitz levels to 6.

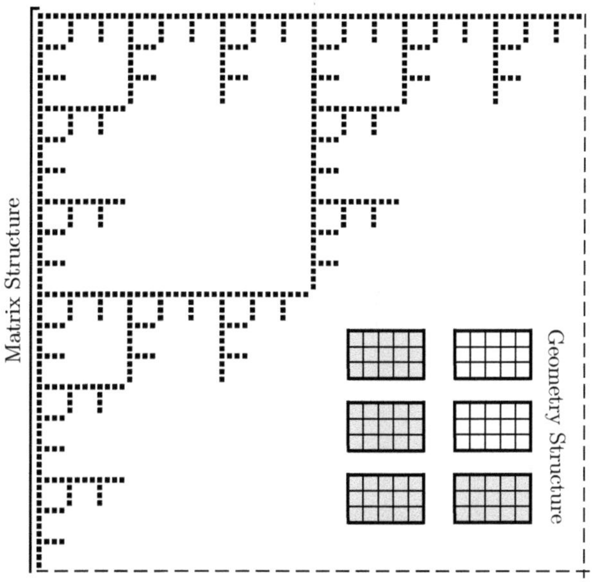

Figure 4.3: Location of the unique entries in $\hat{\mathbf{Z}}_{P\times P}$ when the mesh structure associated with Fig. 4.1 is repeated 3 and 2 times along different directions; The periodicity orders ($n_x = 2$, $n_y = 3$) are visible at the outer (third and fourth) Toeplitz levels. As many as the already picked entries in $\bar{\mathrm{Z}}_\nu$ (the number of unique elements in the lower triangle part of the fictitious square box anchored to the level-border corner) intermediate zeros have to be inserted in $\bar{\underline{\mathrm{I}}}_j$ once the up-right moving selection pointer jumps into another Toeplitz level.

4.4 Toeplitz Property on Time (Order) Indices

The following constellation illustrates the general structure of the system of matrix-vector products in the (A)MOD (3.102) or FDDM (3.130) schemes.

$$
\begin{bmatrix}
\bar{\bar{\mathbf{Z}}}_0\bar{\mathbf{I}}_0\\
\bar{\bar{\mathbf{Z}}}_0\bar{\mathbf{I}}_1\\
\bar{\bar{\mathbf{Z}}}_0\bar{\mathbf{I}}_2\\
\bar{\bar{\mathbf{Z}}}_0\bar{\mathbf{I}}_3\\
\bar{\bar{\mathbf{Z}}}_0\bar{\mathbf{I}}_4\\
\bar{\bar{\mathbf{Z}}}_0\bar{\mathbf{I}}_5\\
\bar{\bar{\mathbf{Z}}}_0\bar{\mathbf{I}}_6\\
\bar{\bar{\mathbf{Z}}}_0\bar{\mathbf{I}}_7\\
\bar{\bar{\mathbf{Z}}}_0\bar{\mathbf{I}}_8\\
\bar{\bar{\mathbf{Z}}}_0\bar{\mathbf{I}}_9\\
\bar{\bar{\mathbf{Z}}}_0\bar{\mathbf{I}}_{10}\\
\bar{\bar{\mathbf{Z}}}_0\bar{\mathbf{I}}_{11}\\
\bar{\bar{\mathbf{Z}}}_0\bar{\mathbf{I}}_{12}\\
\bar{\bar{\mathbf{Z}}}_0\bar{\mathbf{I}}_{13}\\
\bar{\bar{\mathbf{Z}}}_0\bar{\mathbf{I}}_{14}\\
\bar{\bar{\mathbf{Z}}}_0\bar{\mathbf{I}}_{15}\\
\vdots
\end{bmatrix}
=
\begin{bmatrix}
\bar{\mathbf{V}}_0\\ \bar{\mathbf{V}}_1\\ \bar{\mathbf{V}}_2\\ \bar{\mathbf{V}}_3\\ \bar{\mathbf{V}}_4\\ \bar{\mathbf{V}}_5\\ \bar{\mathbf{V}}_6\\ \bar{\mathbf{V}}_7\\ \bar{\mathbf{V}}_8\\ \bar{\mathbf{V}}_9\\ \bar{\mathbf{V}}_{10}\\ \bar{\mathbf{V}}_{11}\\ \bar{\mathbf{V}}_{12}\\ \bar{\mathbf{V}}_{13}\\ \bar{\mathbf{V}}_{14}\\ \bar{\mathbf{V}}_{15}\\ \vdots
\end{bmatrix}
-
\begin{bmatrix}
\bar{\bar{0}} \\
\bar{\bar{\mathbf{Z}}}_1 & \bar{\bar{0}} \\
\bar{\bar{\mathbf{Z}}}_2 & \bar{\bar{\mathbf{Z}}}_1 & \bar{\bar{0}} \\
\bar{\bar{\mathbf{Z}}}_3 & \bar{\bar{\mathbf{Z}}}_2 & \bar{\bar{\mathbf{Z}}}_1 & \bar{\bar{0}} \\
\bar{\bar{\mathbf{Z}}}_4 & \bar{\bar{\mathbf{Z}}}_3 & \bar{\bar{\mathbf{Z}}}_2 & \bar{\bar{\mathbf{Z}}}_1 & \bar{\bar{0}} \\
\bar{\bar{\mathbf{Z}}}_5 & \bar{\bar{\mathbf{Z}}}_4 & \bar{\bar{\mathbf{Z}}}_3 & \bar{\bar{\mathbf{Z}}}_2 & \bar{\bar{\mathbf{Z}}}_1 & \bar{\bar{0}} & & & & & & \underline{\underline{0}} \\
\bar{\bar{\mathbf{Z}}}_6 & \bar{\bar{\mathbf{Z}}}_5 & \bar{\bar{\mathbf{Z}}}_4 & \bar{\bar{\mathbf{Z}}}_3 & \bar{\bar{\mathbf{Z}}}_2 & \bar{\bar{\mathbf{Z}}}_1 & \bar{\bar{0}} \\
\bar{\bar{\mathbf{Z}}}_7 & \bar{\bar{\mathbf{Z}}}_6 & \bar{\bar{\mathbf{Z}}}_5 & \bar{\bar{\mathbf{Z}}}_4 & \bar{\bar{\mathbf{Z}}}_3 & \bar{\bar{\mathbf{Z}}}_2 & \bar{\bar{\mathbf{Z}}}_1 & \bar{\bar{0}} \\
\bar{\bar{\mathbf{Z}}}_8 & \bar{\bar{\mathbf{Z}}}_7 & \bar{\bar{\mathbf{Z}}}_6 & \bar{\bar{\mathbf{Z}}}_5 & \bar{\bar{\mathbf{Z}}}_4 & \bar{\bar{\mathbf{Z}}}_3 & \bar{\bar{\mathbf{Z}}}_2 & \bar{\bar{\mathbf{Z}}}_1 & \bar{\bar{0}} \\
\bar{\bar{\mathbf{Z}}}_9 & \bar{\bar{\mathbf{Z}}}_8 & \bar{\bar{\mathbf{Z}}}_7 & \bar{\bar{\mathbf{Z}}}_6 & \bar{\bar{\mathbf{Z}}}_5 & \bar{\bar{\mathbf{Z}}}_4 & \bar{\bar{\mathbf{Z}}}_3 & \bar{\bar{\mathbf{Z}}}_2 & \bar{\bar{\mathbf{Z}}}_1 & \bar{\bar{0}} \\
\bar{\bar{\mathbf{Z}}}_{10} & \bar{\bar{\mathbf{Z}}}_9 & \bar{\bar{\mathbf{Z}}}_8 & \bar{\bar{\mathbf{Z}}}_7 & \bar{\bar{\mathbf{Z}}}_6 & \bar{\bar{\mathbf{Z}}}_5 & \bar{\bar{\mathbf{Z}}}_4 & \bar{\bar{\mathbf{Z}}}_3 & \bar{\bar{\mathbf{Z}}}_2 & \bar{\bar{\mathbf{Z}}}_1 & \bar{\bar{0}} \\
\bar{\bar{\mathbf{Z}}}_{11} & \bar{\bar{\mathbf{Z}}}_{10} & \bar{\bar{\mathbf{Z}}}_9 & \bar{\bar{\mathbf{Z}}}_8 & \bar{\bar{\mathbf{Z}}}_7 & \bar{\bar{\mathbf{Z}}}_6 & \bar{\bar{\mathbf{Z}}}_5 & \bar{\bar{\mathbf{Z}}}_4 & \bar{\bar{\mathbf{Z}}}_3 & \bar{\bar{\mathbf{Z}}}_2 & \bar{\bar{\mathbf{Z}}}_1 & \bar{\bar{0}} \\
\bar{\bar{\mathbf{Z}}}_{12} & \bar{\bar{\mathbf{Z}}}_{11} & \bar{\bar{\mathbf{Z}}}_{10} & \bar{\bar{\mathbf{Z}}}_9 & \bar{\bar{\mathbf{Z}}}_8 & \bar{\bar{\mathbf{Z}}}_7 & \bar{\bar{\mathbf{Z}}}_6 & \bar{\bar{\mathbf{Z}}}_5 & \bar{\bar{\mathbf{Z}}}_4 & \bar{\bar{\mathbf{Z}}}_3 & \bar{\bar{\mathbf{Z}}}_2 & \bar{\bar{\mathbf{Z}}}_1 & \bar{\bar{0}} \\
\bar{\bar{\mathbf{Z}}}_{13} & \bar{\bar{\mathbf{Z}}}_{12} & \bar{\bar{\mathbf{Z}}}_{11} & \bar{\bar{\mathbf{Z}}}_{10} & \bar{\bar{\mathbf{Z}}}_9 & \bar{\bar{\mathbf{Z}}}_8 & \bar{\bar{\mathbf{Z}}}_7 & \bar{\bar{\mathbf{Z}}}_6 & \bar{\bar{\mathbf{Z}}}_5 & \bar{\bar{\mathbf{Z}}}_4 & \bar{\bar{\mathbf{Z}}}_3 & \bar{\bar{\mathbf{Z}}}_2 & \bar{\bar{\mathbf{Z}}}_1 & \bar{\bar{0}} \\
\bar{\bar{\mathbf{Z}}}_{14} & \bar{\bar{\mathbf{Z}}}_{13} & \bar{\bar{\mathbf{Z}}}_{12} & \bar{\bar{\mathbf{Z}}}_{11} & \bar{\bar{\mathbf{Z}}}_{10} & \bar{\bar{\mathbf{Z}}}_9 & \bar{\bar{\mathbf{Z}}}_8 & \bar{\bar{\mathbf{Z}}}_7 & \bar{\bar{\mathbf{Z}}}_6 & \bar{\bar{\mathbf{Z}}}_5 & \bar{\bar{\mathbf{Z}}}_4 & \bar{\bar{\mathbf{Z}}}_3 & \bar{\bar{\mathbf{Z}}}_2 & \bar{\bar{\mathbf{Z}}}_1 & \bar{\bar{0}} \\
\bar{\bar{\mathbf{Z}}}_{15} & \bar{\bar{\mathbf{Z}}}_{14} & \bar{\bar{\mathbf{Z}}}_{13} & \bar{\bar{\mathbf{Z}}}_{12} & \bar{\bar{\mathbf{Z}}}_{11} & \bar{\bar{\mathbf{Z}}}_{10} & \bar{\bar{\mathbf{Z}}}_9 & \bar{\bar{\mathbf{Z}}}_8 & \bar{\bar{\mathbf{Z}}}_7 & \bar{\bar{\mathbf{Z}}}_6 & \bar{\bar{\mathbf{Z}}}_5 & \bar{\bar{\mathbf{Z}}}_4 & \bar{\bar{\mathbf{Z}}}_3 & \bar{\bar{\mathbf{Z}}}_2 & \bar{\bar{\mathbf{Z}}}_1 & \bar{\bar{0}} \\
 & & & & & & & & & & & & & & & & \ddots
\end{bmatrix}
\begin{bmatrix}
\bar{\mathbf{I}}_0\\ \bar{\mathbf{I}}_1\\ \bar{\mathbf{I}}_2\\ \bar{\mathbf{I}}_3\\ \bar{\mathbf{I}}_4\\ \bar{\mathbf{I}}_5\\ \bar{\mathbf{I}}_6\\ \bar{\mathbf{I}}_7\\ \bar{\mathbf{I}}_8\\ \bar{\mathbf{I}}_9\\ \bar{\mathbf{I}}_{10}\\ \bar{\mathbf{I}}_{11}\\ \bar{\mathbf{I}}_{12}\\ \bar{\mathbf{I}}_{13}\\ \bar{\mathbf{I}}_{14}\\ \bar{\mathbf{I}}_{15}\\ \vdots
\end{bmatrix}
\tag{4.8}
$$

It also represents that of the MOT (3.19) recipes, since $\bar{\mathrm{I}}_0 = 0$ in the MOT:

$$
\begin{bmatrix}
\bar{\bar{\mathbf{Z}}}_0\bar{\mathbf{I}}_1\\
\bar{\bar{\mathbf{Z}}}_0\bar{\mathbf{I}}_2\\
\bar{\bar{\mathbf{Z}}}_0\bar{\mathbf{I}}_3\\
\bar{\bar{\mathbf{Z}}}_0\bar{\mathbf{I}}_4\\
\bar{\bar{\mathbf{Z}}}_0\bar{\mathbf{I}}_5\\
\bar{\bar{\mathbf{Z}}}_0\bar{\mathbf{I}}_6\\
\bar{\bar{\mathbf{Z}}}_0\bar{\mathbf{I}}_7\\
\bar{\bar{\mathbf{Z}}}_0\bar{\mathbf{I}}_8\\
\bar{\bar{\mathbf{Z}}}_0\bar{\mathbf{I}}_9\\
\bar{\bar{\mathbf{Z}}}_0\bar{\mathbf{I}}_{10}\\
\bar{\bar{\mathbf{Z}}}_0\bar{\mathbf{I}}_{11}\\
\bar{\bar{\mathbf{Z}}}_0\bar{\mathbf{I}}_{12}\\
\bar{\bar{\mathbf{Z}}}_0\bar{\mathbf{I}}_{13}\\
\bar{\bar{\mathbf{Z}}}_0\bar{\mathbf{I}}_{14}\\
\bar{\bar{\mathbf{Z}}}_0\bar{\mathbf{I}}_{15}\\
\bar{\bar{\mathbf{Z}}}_0\bar{\mathbf{I}}_{16}\\
\vdots
\end{bmatrix}
=
\begin{bmatrix}
\bar{\mathbf{V}}_1\\ \bar{\mathbf{V}}_2\\ \bar{\mathbf{V}}_3\\ \bar{\mathbf{V}}_4\\ \bar{\mathbf{V}}_5\\ \bar{\mathbf{V}}_6\\ \bar{\mathbf{V}}_7\\ \bar{\mathbf{V}}_8\\ \bar{\mathbf{V}}_9\\ \bar{\mathbf{V}}_{10}\\ \bar{\mathbf{V}}_{11}\\ \bar{\mathbf{V}}_{12}\\ \bar{\mathbf{V}}_{13}\\ \bar{\mathbf{V}}_{14}\\ \bar{\mathbf{V}}_{15}\\ \bar{\mathbf{V}}_{16}\\ \vdots
\end{bmatrix}
-
\begin{bmatrix}
\bar{\bar{0}} \\
\bar{\bar{\mathbf{Z}}}_1 & \bar{\bar{0}} \\
\bar{\bar{\mathbf{Z}}}_2 & \bar{\bar{\mathbf{Z}}}_1 & \bar{\bar{0}} \\
\bar{\bar{\mathbf{Z}}}_3 & \bar{\bar{\mathbf{Z}}}_2 & \bar{\bar{\mathbf{Z}}}_1 & \bar{\bar{0}} \\
\bar{\bar{\mathbf{Z}}}_4 & \bar{\bar{\mathbf{Z}}}_3 & \bar{\bar{\mathbf{Z}}}_2 & \bar{\bar{\mathbf{Z}}}_1 & \bar{\bar{0}} \\
\bar{\bar{\mathbf{Z}}}_5 & \bar{\bar{\mathbf{Z}}}_4 & \bar{\bar{\mathbf{Z}}}_3 & \bar{\bar{\mathbf{Z}}}_2 & \bar{\bar{\mathbf{Z}}}_1 & \bar{\bar{0}} & & & & & & \underline{\underline{0}} \\
\bar{\bar{\mathbf{Z}}}_6 & \bar{\bar{\mathbf{Z}}}_5 & \bar{\bar{\mathbf{Z}}}_4 & \bar{\bar{\mathbf{Z}}}_3 & \bar{\bar{\mathbf{Z}}}_2 & \bar{\bar{\mathbf{Z}}}_1 & \bar{\bar{0}} \\
\bar{\bar{\mathbf{Z}}}_7 & \bar{\bar{\mathbf{Z}}}_6 & \bar{\bar{\mathbf{Z}}}_5 & \bar{\bar{\mathbf{Z}}}_4 & \bar{\bar{\mathbf{Z}}}_3 & \bar{\bar{\mathbf{Z}}}_2 & \bar{\bar{\mathbf{Z}}}_1 & \bar{\bar{0}} \\
\bar{\bar{\mathbf{Z}}}_8 & \bar{\bar{\mathbf{Z}}}_7 & \bar{\bar{\mathbf{Z}}}_6 & \bar{\bar{\mathbf{Z}}}_5 & \bar{\bar{\mathbf{Z}}}_4 & \bar{\bar{\mathbf{Z}}}_3 & \bar{\bar{\mathbf{Z}}}_2 & \bar{\bar{\mathbf{Z}}}_1 & \bar{\bar{0}} \\
\bar{\bar{\mathbf{Z}}}_9 & \bar{\bar{\mathbf{Z}}}_8 & \bar{\bar{\mathbf{Z}}}_7 & \bar{\bar{\mathbf{Z}}}_6 & \bar{\bar{\mathbf{Z}}}_5 & \bar{\bar{\mathbf{Z}}}_4 & \bar{\bar{\mathbf{Z}}}_3 & \bar{\bar{\mathbf{Z}}}_2 & \bar{\bar{\mathbf{Z}}}_1 & \bar{\bar{0}} \\
\bar{\bar{\mathbf{Z}}}_{10} & \bar{\bar{\mathbf{Z}}}_9 & \bar{\bar{\mathbf{Z}}}_8 & \bar{\bar{\mathbf{Z}}}_7 & \bar{\bar{\mathbf{Z}}}_6 & \bar{\bar{\mathbf{Z}}}_5 & \bar{\bar{\mathbf{Z}}}_4 & \bar{\bar{\mathbf{Z}}}_3 & \bar{\bar{\mathbf{Z}}}_2 & \bar{\bar{\mathbf{Z}}}_1 & \bar{\bar{0}} \\
\bar{\bar{\mathbf{Z}}}_{11} & \bar{\bar{\mathbf{Z}}}_{10} & \bar{\bar{\mathbf{Z}}}_9 & \bar{\bar{\mathbf{Z}}}_8 & \bar{\bar{\mathbf{Z}}}_7 & \bar{\bar{\mathbf{Z}}}_6 & \bar{\bar{\mathbf{Z}}}_5 & \bar{\bar{\mathbf{Z}}}_4 & \bar{\bar{\mathbf{Z}}}_3 & \bar{\bar{\mathbf{Z}}}_2 & \bar{\bar{\mathbf{Z}}}_1 & \bar{\bar{0}} \\
\bar{\bar{\mathbf{Z}}}_{12} & \bar{\bar{\mathbf{Z}}}_{11} & \bar{\bar{\mathbf{Z}}}_{10} & \bar{\bar{\mathbf{Z}}}_9 & \bar{\bar{\mathbf{Z}}}_8 & \bar{\bar{\mathbf{Z}}}_7 & \bar{\bar{\mathbf{Z}}}_6 & \bar{\bar{\mathbf{Z}}}_5 & \bar{\bar{\mathbf{Z}}}_4 & \bar{\bar{\mathbf{Z}}}_3 & \bar{\bar{\mathbf{Z}}}_2 & \bar{\bar{\mathbf{Z}}}_1 & \bar{\bar{0}} \\
\bar{\bar{\mathbf{Z}}}_{13} & \bar{\bar{\mathbf{Z}}}_{12} & \bar{\bar{\mathbf{Z}}}_{11} & \bar{\bar{\mathbf{Z}}}_{10} & \bar{\bar{\mathbf{Z}}}_9 & \bar{\bar{\mathbf{Z}}}_8 & \bar{\bar{\mathbf{Z}}}_7 & \bar{\bar{\mathbf{Z}}}_6 & \bar{\bar{\mathbf{Z}}}_5 & \bar{\bar{\mathbf{Z}}}_4 & \bar{\bar{\mathbf{Z}}}_3 & \bar{\bar{\mathbf{Z}}}_2 & \bar{\bar{\mathbf{Z}}}_1 & \bar{\bar{0}} \\
\bar{\bar{\mathbf{Z}}}_{14} & \bar{\bar{\mathbf{Z}}}_{13} & \bar{\bar{\mathbf{Z}}}_{12} & \bar{\bar{\mathbf{Z}}}_{11} & \bar{\bar{\mathbf{Z}}}_{10} & \bar{\bar{\mathbf{Z}}}_9 & \bar{\bar{\mathbf{Z}}}_8 & \bar{\bar{\mathbf{Z}}}_7 & \bar{\bar{\mathbf{Z}}}_6 & \bar{\bar{\mathbf{Z}}}_5 & \bar{\bar{\mathbf{Z}}}_4 & \bar{\bar{\mathbf{Z}}}_3 & \bar{\bar{\mathbf{Z}}}_2 & \bar{\bar{\mathbf{Z}}}_1 & \bar{\bar{0}} \\
\bar{\bar{\mathbf{Z}}}_{15} & \bar{\bar{\mathbf{Z}}}_{14} & \bar{\bar{\mathbf{Z}}}_{13} & \bar{\bar{\mathbf{Z}}}_{12} & \bar{\bar{\mathbf{Z}}}_{11} & \bar{\bar{\mathbf{Z}}}_{10} & \bar{\bar{\mathbf{Z}}}_9 & \bar{\bar{\mathbf{Z}}}_8 & \bar{\bar{\mathbf{Z}}}_7 & \bar{\bar{\mathbf{Z}}}_6 & \bar{\bar{\mathbf{Z}}}_5 & \bar{\bar{\mathbf{Z}}}_4 & \bar{\bar{\mathbf{Z}}}_3 & \bar{\bar{\mathbf{Z}}}_2 & \bar{\bar{\mathbf{Z}}}_1 & \bar{\bar{0}} \\
 & & & & & & & & & & & & & & & & \ddots
\end{bmatrix}
\begin{bmatrix}
\bar{\mathbf{I}}_1\\ \bar{\mathbf{I}}_2\\ \bar{\mathbf{I}}_3\\ \bar{\mathbf{I}}_4\\ \bar{\mathbf{I}}_5\\ \bar{\mathbf{I}}_6\\ \bar{\mathbf{I}}_7\\ \bar{\mathbf{I}}_8\\ \bar{\mathbf{I}}_9\\ \bar{\mathbf{I}}_{10}\\ \bar{\mathbf{I}}_{11}\\ \bar{\mathbf{I}}_{12}\\ \bar{\mathbf{I}}_{13}\\ \bar{\mathbf{I}}_{14}\\ \bar{\mathbf{I}}_{15}\\ \bar{\mathbf{I}}_{16}\\ \vdots
\end{bmatrix}
\tag{4.9}
$$

These systems of equations is not to be solved all at once; rather, it is solved recursively. First, the current for the first time order (row) is found. It is then used to find the current at the next higher time (order) step and so forth. All the matrix-vector multiplications corresponding to the row i in (4.9) or n in (4.8) are needed to be computed in order to construct the right side of (3.102), (3.130), or (3.19) for the increment of order i in the MOD or FDDM (or time step n in the MOT). As a consequence of causality and the use of the retarded Green's function as well as bases orthogonality (3.64) ($I_\nu = 0 \quad \nu < 0$), the polynomial system matrix is block lower triangular on time (order) indices. In fact, the MOT, MOD, and FDDM schemes are all block forward-substitution solution methods based on the block lower triangularity of the system matrix. The system matrix may become band diagonal when the lower bound of summation in (3.19) modifies to $r = \max(1, n - N_g)$ as explained in (3.22). The system matrix is also (block) Toeplitz on the time (order) stairs, as a consequence of invariance under time (order) translations, e.g. all schemes in Section 3.4.4 are always function of $\nu = i - j$ and not i or j alone. The Toeplitz property on the temporal dimension is immediately visible in Fig. 4.4. Specifically, row n in Fig. 4.4 represents the products that have to be concatenated to the excitation vector $\bar{\mathrm{V}}$ at time step n. In what follows, instead of direct row-by-row multiplication, efficient mechanisms are investigated for building the right side of (3.102), (3.130), or (3.19).

The FFTs have been used to accelerate Toeplitz matrix-vector multiplications representing spatial convolutions on a uniform grid in Section 4.2. Due to the time-stepping nature of the MOT, MOD, and FDDM solution procedures, however, the spatial FFT can not be used in the same manner to accelerate the temporal convolutions. When the solver operates within an marching framework, the current vectors are unknown at future time steps. Filling the future unknown current vectors with zeros to compute the multiplication at time step n is clearly inefficient, since a similar procedure would have to be repeated each and every time step. Fig. 4.4 elucidates the mechanism of grouping in computation of the matrix-vector multiply for the first 16 time order steps to take advantages of the temporal Toeplitz structure. The square groups of matrices $\bar{\bar{\mathrm{Z}}}_\nu$ are block-Toeplitz on two levels (i.e., the aggregates of Toeplitz blocks are repeated) which can benefit their multiplications by the current vectors without violating causality. That is, only currents that are already known are required to multiply the interaction matrices when marching in time or order ranks. The aggregates of the individual matrix blocks indicated by the superimposed square in Fig. 4.4 are referred to as block aggregates. Obviously, the block arrangement introduced for aggregate matrix-vector multiply in the MOT can be shifted one block up to well match in to the MOD or FDDM system matrix structures.

Assume that the current vectors up to the time (order) step n have been already calculated and we are about to fill the right side of (3.19) and solve it for the next time step $n + 1$. Since the aggregates of matrix blocks are themselves block-Toeplitz, the computation process of matrix-vector multiplications is partitioned into convolving dyadic blocks indicated by squares in Fig. 4.4. The alignment of the blocks and the arrangement of the auxiliary arrays elements are all to be addressed by the binary representation of $n|_b = b_{\lfloor \log n \rfloor} \cdots b_k \cdots b_2 b_1 b_0$, where $|_b$ expresses an integer in the binary format. In transition from time step n to the one higher, the kth bit of $n|_b$ turns from 0 into 1 and the scheme calls for the multiplication of the new block aggregate of size $\alpha(M)(B_k \times B_k)$, where $B_k = 2^k$. To better perceive the linkage of the algorithm, first the multiplication of the most left blocks is inspected. Vectors $\bar{\mathrm{X}}_k$ are first calculated whenever $n = 2^k$ (k is a

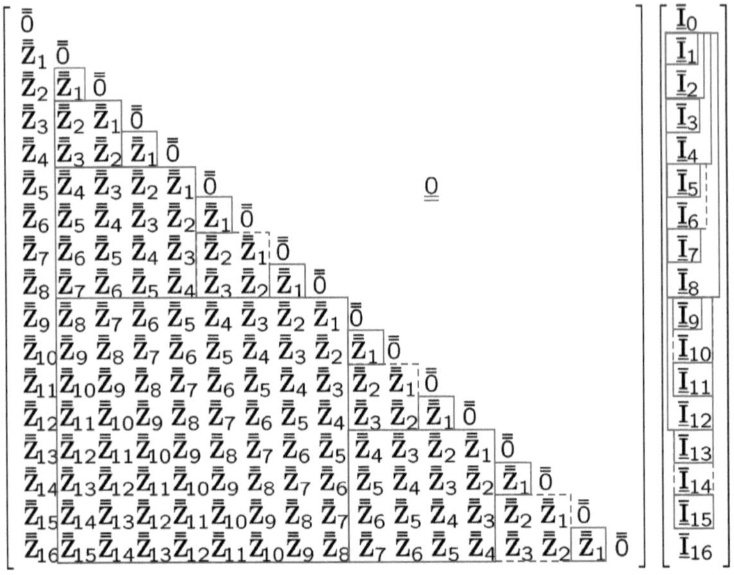

Figure 4.4: The beginning 17 order chains of the (A)MOD (3.102) or the FDDM (3.130) algorithms for $i = 0, 1, 2, \ldots, 16$ or equivalently, when $\bar{\mathrm{I}}_0 = \bar{\bar{0}}$, the early 16 time steps of the MOT (3.19) $n = 1, 2, \ldots, 16$.

natural number),

$$\begin{pmatrix} \bar{\mathrm{X}}_k(1) \\ \bar{\mathrm{X}}_k(2) \\ \vdots \\ \bar{\mathrm{X}}_k(n) \end{pmatrix} = \begin{pmatrix} \bar{\bar{\mathbf{Z}}}_n & \bar{\bar{\mathbf{Z}}}_{n-1} & \cdots & \bar{\bar{\mathbf{Z}}}_1 \\ \bar{\bar{\mathbf{Z}}}_{n+1} & \bar{\bar{\mathbf{Z}}}_n & \cdots & \bar{\bar{\mathbf{Z}}}_2 \\ \vdots & \vdots & \ddots & \vdots \\ \bar{\bar{\mathbf{Z}}}_{2n-1} & \bar{\bar{\mathbf{Z}}}_{2n-2} & \cdots & \bar{\bar{\mathbf{Z}}}_n \end{pmatrix} \begin{pmatrix} \bar{\mathbf{I}}_1 \\ \bar{\mathbf{I}}_2 \\ \vdots \\ \bar{\mathbf{I}}_n \end{pmatrix}. \tag{4.10}$$

Note that the underlines in Fig. 4.4 indicate that the current vectors may have been already zero-padded for accelerating space convolutions on uniform grids by the space-FFT, Section 4.2. The convolution product (4.10) are obtained from element-by-element multiplication of the transformed version of

$$\begin{aligned} \bar{\mathrm{Z}}_k &= \left[\bar{\bar{\mathbf{Z}}}_{2n-1}\, \bar{\bar{\mathbf{Z}}}_{2n-2} \cdots \bar{\bar{\mathbf{Z}}}_1\right] \\ \underline{\bar{\mathrm{I}}}_n^k &= \left[\bar{\mathbf{I}}_n \;\cdots\; \bar{\mathbf{I}}_2\; \bar{\mathbf{I}}_1\; 0 \;\cdots\; 0\right]_{1\times 2n-1} \end{aligned}$$

in spectral frequency domain and this is the time when the $\mathrm{FFT}\{\bar{\mathrm{Z}}_k\}$ are stored for further usage on the next 2×2^kth steps. Collectively, the convolution matrix-vector product on the right side of (3.102), (3.130), or (3.19), thus, can be computed by

$$\sum_{j=0}^{i-1} \bar{\mathrm{Z}}_\nu \bar{\mathrm{I}}_j \Big\rfloor_{\text{(A)MOD}} \equiv \sum_{j=0}^{i-1} \bar{\bar{\mathrm{Z}}}_{i-j} \bar{\mathrm{I}}_j \Big\rfloor_{\text{FDDM}} \equiv \sum_{r=1}^{n-1} \bar{\bar{\mathrm{Z}}}_{n-r} \bar{\mathrm{I}}_r \Big\rfloor_{\text{MOT}} = \sum_{k=0}^{\lfloor \log n \rfloor} b_k\; \bar{\mathrm{X}}_k(\kappa + 1) \tag{4.11}$$

where the sign $\equiv$ reads as corresponds to, and the integer $\kappa|_b = b_{k-1}\cdots b_1 b_0$ is assigned by truncation of the first k digits of n in its binary format and the auxiliary vectors $\bar{\mathrm{X}}_k$ with length $B_k = 2^k$ are defined as

$$
\begin{aligned}
[\bar{\mathrm{X}}_k]_{B_k\times 1} &= \bar{\mathrm{Z}}_k \otimes \bar{\underline{\mathrm{I}}}_n^k \\
&= \dagger\mathrm{Re}\{\mathrm{FFT}^{-1}\{\mathrm{FFT}\{\bar{\mathrm{Z}}_k\}\cdot\mathrm{FFT}\{\bar{\underline{\mathrm{I}}}_n\}\}\} \\
\bar{\mathrm{Z}}_k &= \left[\bar{\bar{\mathbf{Z}}}_{2B_k-1}\ \bar{\bar{\mathbf{Z}}}_{2B_k-2}\ \cdots\ \bar{\bar{\mathbf{Z}}}_1\right] \\
\bar{\underline{\mathrm{I}}}_n^k &= \left[\bar{\mathbf{I}}_n\ \bar{\mathbf{I}}_{n-1}\ \cdots\ \bar{\mathbf{I}}_{n-(B_k-1)}\ 0\ \cdots\ 0\right]_{1\times 2B_k-1}
\end{aligned}
\tag{4.12}
$$

in which $\bar{\mathrm{Z}}_k$ is a vector consisting of the unique entries in the kth individual block of the matrix aggregates, encompassed by solid lines in Fig. 4.4, and the vector $\bar{\underline{\mathrm{I}}}_n^k$ is the flipped up/down and zero-padded extension of the behind B_k current sequences with the same size as $\bar{\mathrm{Z}}_k$. The operator $\dagger$ only extracts the desired array elements corresponding to the positions of the nonzero $\bar{\mathrm{I}}_r$ entries in $\bar{\underline{\mathrm{I}}}_n^k$. Namely, $\dagger\mathrm{Re}\{\}$ flips the resulting sequence in down/up direction and picks up the real part of the first B_k original elements. The summation operation is skipped in (4.11) when the flag b_k is off, and as a result, at each time (order) step only one of the blocked aggregates, indicated by squares in Fig. 4.4, is multiplied. The vector $\bar{\mathrm{X}}_k$ in (4.12) is first calculated on the time step $n = B_k$ and then updated whenever the kth bit in binary representation of $n|_b$ changes from 0 to 1 in transition to the next time (order) step, which happens every $2B_k$ cycles. Accordingly, every block is multiplied once every $2B_k$ time (order) steps. At each time step, only recent current coefficients are transformed to the spectral-frequency domain, and after element-by-element multiplication, the auxiliary vectors are inverse transformed to accumulate the future scattered fields. The future retarded fields are constructed partially at each time step in $\bar{\mathrm{X}}_k$, and the complete scattered field at a given time step is available only at that time step. For instance, at time step fourteen ($13|_b = 1101$) (4.11) includes $\bar{\mathrm{X}}_3(6)+\bar{\mathrm{X}}_2(2)+\bar{\mathrm{X}}_0(1)$. Causality is not violated since the past (known) current vectors are only used to find the present and future fields.

The time-FFT routine (4.12) can be amalgamated with the space-FFT associated with the Toeplitz property of individual interaction matrices $\bar{\bar{\mathrm{Z}}}_\nu$ in Section 4.2, $\alpha(M) = M\log M$. They can be merged into single 1-D FFT, once the encompassing Toeplitz property on the time-order dimension are placed above the nested Toepliz levels owing to uniform meshing, as Section 4.3 adjoins the space-periodicity. When the unification of space-time FFT are not used, for small groups of Toeplitz matrix-vector multiply, direct computations are faster than the time-FFT approach. Hence, small groups of the early time matrices can be kept out from the FFT procedure on the aggregates. In other words, for blocks smaller than an elementary size B, enclosed by the lower triangular boxes in Fig. 4.5, the matrix-vector multiply is performed conventionally without using FFT [99]. As an alternative approach for fast evaluation of the matrix-vector multiply, varying-size blocks larger than $B\times B$ in Fig. 4.4 are interchanged by (subdivided into) aggregates of elementary subblocks, specified by the squares superimposed on Fig. 4.5. In this case, the square block sizes in (4.12) and the upper limit of $\sum_{k=0}$ in (4.11) are scaled to $B_k = B$ and $\lfloor\log n\rfloor/B$, respectively. When this fixed-size grouping is incorporated in the spatial FFT routines, however, the effective size of the elementary triangular shadow boxes shrinks and for large M it vanishes entirely. Note that due to the causality, the lower triangular blocks in Fig. 4.5 can not be considered as square boxes half filled by zeros. Local interpolation/extrapolations can

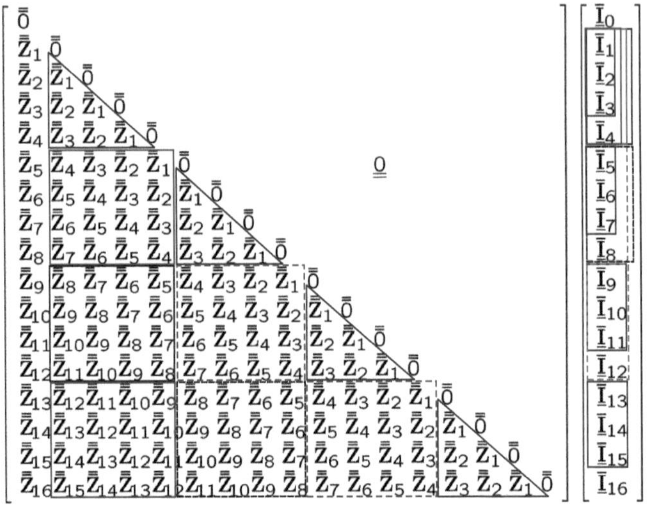

Figure 4.5: Aggregates of fixed-size Toeplitz blocks, $B = 4$.

adopt the FFT routines to nonequidistantly temporal grids [102], so that they can be applied for nonuniform time stepping as well.

4.4.1 Computational Complexity Analysis

The transformed, compressed version (unique entries) of $\bar{\bar{Z}}_\nu$ matrices, FFT$\{\bar{Z}_\nu\}$ with dimensions $\mathcal{O}(M)$, are stored in memory for further usage in constructing (3.102), (3.130), or (3.19) and solving them for next higher orders of i. Direct computation of the aggreate matrix vector multiply will require $\alpha(M)N^2$ operations. Since the discrete Fourier transforms of matrix aggregates are frequently used throughout the marching process, they are computed and stored in advance, and thus, they are omitted from the cost computation. The computational expenses of performing the matrix-vector multiply using FFT for a $B_k \times B_k$ Toeplitz matrix block and the zero-padded recent current vector of size $B_k \times 1$ involves $\mathcal{O}(B_k \log B_k)$ operations, since it includes, approximately

- $(2B_k - 1)\log(2B_k - 1)$ operations to transform the current vector to the spectral-frequency domain

- $(2B_k - 1)$ CPU cycles to multiply the unique entries of the blocked aggregate with the deformed current vector in frequency domain

- $(2B_k - 1)\log(2B_k - 1)$ costs the inverse transformation of the product from the spectral domain back to the spatial domain.

Therefore, the overall complexity is $\mathcal{O}(B_k \log B_k)$.

The stored FFT sequences of individual blocks are consecutively called every $2B_k$ time steps to contribute in the calculation of the products on the upcoming B_k time steps. The

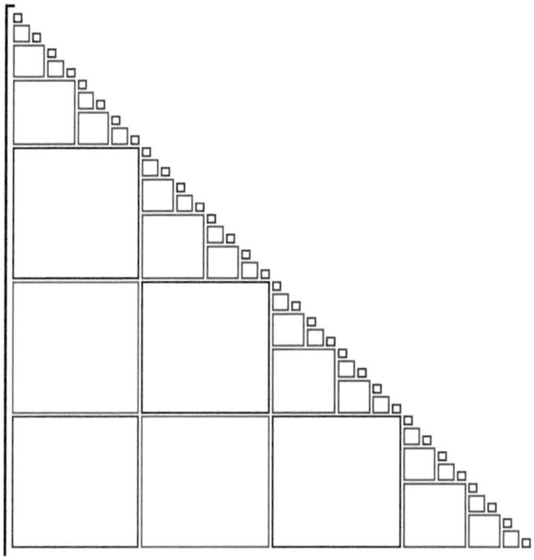

Figure 4.6: Amalgamation of the varying-size and fixed-size aggregates of the retarded interaction matrix blocks.

cost of multiplying the aggregates of matrix blocks with the current vectors varies from time step to time step, i.e., According to Fig. 4.4, the more time (order echelons) passes, the more elements of the current vector are known, allowing faster multiplication of larger blocks via the FFT. As a consequence, the total operation count is

$$\mathcal{O}\left(\alpha(M)\left(\frac{N}{2}\log\frac{N}{2}+2\left(\frac{N}{4}\log\frac{N}{4}\right)+4\left(\frac{N}{8}\log\frac{N}{8}\right)+\cdots+\frac{N}{2}\right)\right)$$
$$=\mathcal{O}\left(\alpha(M)\sum_{k=1}^{\lfloor\log N\rfloor}\frac{N}{2B_k}B_k\log B_k\right)=\mathcal{O}\big(\alpha(M)N\log^2 N\big)\,. \tag{4.13}$$

To avoid any ambiguity, it is worth noting that although the large Toeplitz block can be partitioned further into groups of Toeplitz subblocks, apparently considering higher nested redundant levels for the time Toeplitz property inside the square block-aggregates does not benefit additional computational efficiency. When the larger block aggregates are superseded into aggregates of smaller subblocks with size B, once the reminder of division of n by B is zero, the newly appeared most left block and the corresponding group of the latest current vectors are needed to be transferred to the spectral domain, the rest have been already calculated and stored in the past every B steps. In the Fig. 4.5 strategy, thus, every B time steps a new block is appended whose FFT is needed to be calculated and stored so as to be reused in the next B steps. Therefore, the algorithm calls for $\alpha(M)$ times

$$\mathcal{O}\left(\sum_{k=1}^{\frac{N}{B}-1}(kB+B\log B)\right)=\mathcal{O}\left(\frac{N^2}{B}+N\log B\right) \tag{4.14}$$

operations to compute all the convolutions required for N time (order) steps of simulation. The first term $\mathcal{O}\left(\frac{N^2}{B}\right)$ is related to the vector inner products in spectral domain and it dominates if $B \ll N$. In such circumstances as the algorithm loses its multilevel nature, the running time increases unfavorably. Considering the numbers of individual equally-sized boxes $n_B = \frac{N}{B}$, in order to make the second logarithmic term, relating to the FFT transformations, dominant in comparison with the first term $\mathcal{O}(n_B N)$, one has to choose $n_B \ll N$, which is fulfilled in the case of large box sizes B. When a large B is opted, however, one has left huge portions of corner triangular boxes for costly direct conventional multiplications. Thus, a compromise can be reached by the combination of Fig. 4.4 and Fig. 4.5 configurations as sketched in Fig. 4.6. This means that as long as the kth bit of $n|_b$ satisfies $k < n_B$, multi-size blocks take part in (4.11), otherwise the fixed-size blocks are appointed. This implies that right after every 2^{n_B} time steps (n_B is the number of different multi-size blocks), only the contemporary largest block with the fixed-size B is transformed and so forth. To combine the bounding time FFT and the nested space FFT routines according to the [97] zero sequences have to be inserted into and in between $\bar{\mathrm{I}}_n$, yielding $\bar{\mathrm{I}}_n^k$ for proper alignment based on separation between temporal subblocks (spatial blocks). The storage of the (transformed) compressed unique blocks demands $\mathcal{O}(NM)$ memory, and without the AIM fusion $\mathcal{O}(NM^2)$.

4.5 Wavelet-Based Matrix Compression

The hierarchical bases have been used to solve the TDIE by the MOT methods [103]. Instead of direct multiplication of the dense MOD or FDDM interaction matrices with past current vectors, this section examines the discrete fast wavelet (packet) transform (FWT) to benefit sparse individual matrix-vector multiply for the early slowly-varying kernels grouped by triangular boxes in Fig. 4.6. The use of FWT weakens the mutual interaction of non-overlapping bases and diminishes strong correlation among the expansion coefficients [104]. This is due to numerous useful features of wavelets, including the natural support for multiresolution analysis, the localization in both the spatial and spectral domains and the vanishing moment properties [105]. As a result, the use of FWT in BEM can lead to sparse matrices after thresholding. The interaction matrices are transferred from space domain in 3.130 to a multiresolution wavelet domain by using the FWT, i.e.,

$$\bar{\bar{\mathrm{Z}}}_0 \bar{\mathrm{I}}_i = \bar{\mathrm{V}}_i - \bar{\gamma}_{i-1} - \bar{\beta}_i \qquad i = 0, 1, 2, \ldots, N, \tag{4.15}$$

$$\bar{\beta}_i = \sum_{j=0}^{i-1} \bar{\bar{\mathrm{Z}}}_\nu \bar{\mathrm{I}}_j \simeq \mathrm{IFWT}\{\sum_{j=0}^{i-1} \dagger\mathrm{FWT}\{\bar{\bar{\mathrm{Z}}}_\nu\} \cdot \mathrm{FWT}\{\bar{\mathrm{I}}_j\}\} \tag{4.16}$$

where the † represents hard thresholding and the dot implies that matrix-vector multiplication is carried out in sparse form and the IFWT denotes the inverse FWT. The transformed interaction matrices contains negligible elements which can be omitted without affecting the quality of approximate solution when the thresholding level is chosen carefully [106] based on the norm of the impedance matrix.

4.5.1 Wavelet Packet Transform

Basic wavelet theory can be found in [105]. Here, we briefly review the FWT that are used in matrix compression. Let $\phi(x)$ and $\psi(x)$ be the scaling function and the corresponding mother wavelet function, respectively. They satisfy the well-known two-scale relations,

$$\begin{aligned}\phi(x) &= \sum_k h_k \phi(2x-k) \\ \psi(x) &= \sum_k g_k \psi(2x-k)\end{aligned} \tag{4.17}$$

where h_k and g_k are, respectively, the impulse responses of two quadrature mirror filters, i.e. the low-pass and high-pass filters coefficients in decomposition part of a two channel filter bank structure [107].

Let us now consider a discretized data sequence in the space domain as a vector a^0 of size $M = 2^{J_0}$. The discrete wavelet transform is defined by

$$\begin{aligned}a^{j+1}(n) &= \sum_k h(2n-k)a^j(k) \\ d^{j+1}(n) &= \sum_k g(2n-k)a^j(k)\end{aligned} \tag{4.18}$$

in which the vector a^j at the stage j is decomposed into the smooth approximation part a^{j+1} and the detail (sharply varying) part d^{j+1} by passing the input sequence through the decomposition quadrature filters and then down-sampling the outputs by two (discarding even sample of data). The discrete wavelet "packet" transform (WPT) is a generalization of the commonly used basic wavelet transform (4.18) where the detail part of the vector is decomposed as well. Continuing the process of recursive decomposition of the approximation and detail vectors a^{j+1} and d^{j+1} for $j = 0, \ldots, J-1$, one obtains the full discrete WPT of the vector a^0 at resolution level J.

Conversely, the sequence a^j at stage j can be perfectly reconstructed from the two sequences a^{j+1} and d^{j+1} at stage $j+1$, similarly using the two reconstruction quadrature filters coefficients h'_k (low-pass) and g'_k (high-pass) [107]. The reconstruction by the IFWT is equivalent to first up-sampling the two sequences by two (inserting a zero between each two data samples) and passing the resulting sequence through the two reconstruction quadrature filters and then summing the two outputs, i.e.,

$$a^j(n) = \sum_k h'(2n-k)a^{j+1}(k) + \sum_k g'(2n-k)d^{j+1}(k)$$

Considering the rows and columns of the interaction matrices $\bar{\bar{Z}}_\nu$ as space domain data sampled with the finest spatial resolution, the best suited decomposition tree structure is sought based on the least cost (energy of the transformed matrix). Associating the space samples to the top node of the decomposition tree (that is the finest spatial resolution level), a decision whether to go on decomposing any tree nodes (stepping down to a coarse resolution level) is made based on the comparison of the values of the cost function at the parent and children nodes. If the value of cost at the parent node is greater than its value at its two children (i.e., the sparsity of decomposed vector increases) the decomposition is accepted and further decomposition is applied to that branch at the next stage; otherwise the tree is not decomposed further at that node. This does not necessitate the spread of the decomposition tree over the entire $\log_2^N$ successive resolution levels. The same transformation is applied to the corresponding current vectors $\bar{\mathrm{I}}_j$. After multiplication of the transformed late current vectors with already saved early matrices in sparse formats, (4.19) is applied for up-sampling and two-channel filtering to reconstruct the product.

4.6 Adaptive Integral Methods and Precorrected-FFT

The AIM translates the original basis functions to an equivalent sparse uniform grid to utilize the translational invariance property of the integral kernel for storage reduction and Toeplitz matrix-vector products acceleration using the space-FFT. In the AIM, the whole geometry is enclosed in a regular rectangular grid box [108]. Equivalent auxiliary point sources are placed at each node whose strength is assigned by matching multiple moments of the basis functions between the original and uniform grids. The field of each interior edge is represented by (locally projected onto) clusters of delta current sources on the nodes of the uniform gird. The new expansion associated with the kth edge reads as

$$\mathbf{A}_k = \sum_{q=1}^{M^3} \delta(x - x_{kq})\delta(y - y_{kq})\delta(z - z_{kq})[\Lambda_{kq}^x \hat{\mathbf{x}} + \Lambda_{kq}^y \hat{\mathbf{y}} + \Lambda_{kq}^z \hat{\mathbf{z}}] \tag{4.19}$$

and similarly for the surface divergence of the basis functions

$$\phi_k = \sum_{q=1}^{M^3} \delta(x - x_{kq})\delta(y - y_{kq})\delta(z - z_{kq})\Lambda_{kq}^d \tag{4.20}$$

where M is the expansion order, $\delta(x)$ is the Dirac delta function, and $\mathbf{r}_{kq} = (x_{kq}, y_{kq}, z_{kq})$ denote the position vectors of M^3 points on the grid surrounding the center of the edge (x_0, y_o, z_0). The relation between the new equivalence coefficients $\Lambda_k^{x,y,z,d}$ and the original expansion coefficients I_k in (2.33) is determined by equating the moments of the two expansion basis sets up to order M so that the auxiliary point sources generate nearly identical transient far fields to the primary dihedral triangular elements,

$$\begin{aligned}
&\int_{-\infty}^{\infty}\int_{-\infty}^{\infty}\int_{-\infty}^{\infty} \mathbf{A}_k (x - x_0)^{q_x} (y - y_0)^{q_y} (z - z_0)^{q_z} \, \mathrm{d}x \, \mathrm{d}y \, \mathrm{d}z \qquad \text{for } 0 \le q_x, q_y, q_z \le M \\
&= \sum_{q=1}^{M^3} (x_{kq} - x_0)^{q_x} (y_{kq} - y_0)^{q_y} (z_{kq} - z_0)^{q_z} [\Lambda_{kq}^x \hat{\mathbf{x}} + \Lambda_{kq}^y \hat{\mathbf{y}} + \Lambda_{kq}^z \hat{\mathbf{z}}] \quad \text{with } q = q_x + q_y + q_z \\
&\equiv \int_{-\infty}^{\infty}\int_{-\infty}^{\infty}\int_{-\infty}^{\infty} \mathbf{f}_k(\mathbf{r}) (x - x_0)^{q_x} (y - y_0)^{q_y} (z - z_0)^{q_z} \, \mathrm{d}x \, \mathrm{d}y \, \mathrm{d}z
\end{aligned} \tag{4.21}$$

and similarly for the divergence of the basis functions

$$\begin{aligned}
&\int_{-\infty}^{\infty}\int_{-\infty}^{\infty}\int_{-\infty}^{\infty} \phi_k (x - x_0)^{q_x} (y - y_0)^{q_y} (z - z_0)^{q_z} \, \mathrm{d}x \, \mathrm{d}y \, \mathrm{d}z \qquad \text{for } 0 \le q_x, q_y, q_z \le M \\
&= \sum_{q=1}^{M^3} (x_{kq} - x_0)^{q_x} (y_{kq} - y_0)^{q_y} (z_{kq} - z_0)^{q_z} \Lambda_{kq}^d \qquad \text{with } q = q_x + q_y + q_z \\
&\equiv \int_{-\infty}^{\infty}\int_{-\infty}^{\infty}\int_{-\infty}^{\infty} \nabla_{\mathbf{r}} \cdot \mathbf{f}_k(\mathbf{r}) (x - x_0)^{q_x} (y - y_0)^{q_y} (z - z_0)^{q_z} \, \mathrm{d}x \, \mathrm{d}y \, \mathrm{d}z.
\end{aligned} \tag{4.22}$$

These give four $M^3 \times M^3$ linear system of equations for unknowns $\Lambda^x_{kq}, \Lambda^y_{kq}, \Lambda^z_{kq}, \Lambda^d_{kq}$. A fast closed-form recursive scheme to calculate the moments can be found in [109].

The AIM partitions all the impedance matrices $\bar{\bar{Z}}_1$ through $\bar{\bar{Z}}_{N_g}$ (4.11) into exact "near-field" and approximate "far-field" pairs, that is if the distance between centers of two interacting current elements m^{th} and k^{th} is larger than a prespecified distance, they lie in the far-field region of each other. The total near and far interactions are calculated using

$$\sum_{j=0}^{i-1} \bar{\bar{Z}}_\nu \bar{I}_j = \sum_{j=0}^{i-1} \bar{\bar{Z}}_\nu^{\text{near}} \bar{I}_j + \sum_{j=0}^{i-1} \bar{\bar{Z}}_\nu^{\text{FFT}} \bar{I}_j, \tag{4.23}$$

in which

$$\bar{\bar{Z}}_\nu^{\text{FFT}} = \sum_{l=1}^{4} \bar{\bar{\Lambda}}_l \bar{\bar{G}}_\nu \bar{\bar{\Lambda}}_l^{\text{T}} \tag{4.24}$$

where the retarded Toeplitz matrices $\bar{\bar{G}}_\nu$ are constituted of the free-space Green's function values weighted by $I_\nu(s\frac{R}{c})$ at the regular grid points. The matrices $\bar{\bar{\Lambda}}_l$, $l \in \{x, y, z, d\}$ and

$$\bar{\bar{Z}}_\nu^{\text{near}} = \bar{\bar{Z}}_\nu^{\text{near MoM}} - \bar{\bar{Z}}_\nu^{\text{near AIM}} \tag{4.25}$$

are sparse. The pre-correcting matrix $\bar{\bar{Z}}_\nu^{\text{near}}$ in (4.25) modifies the nearby elements interactions in $\bar{\bar{Z}}_\nu^{\text{FFT}}$ wherever needed by the exact near fields due to the primary current elements computed by the conventional MoM. Transient fields produced by the auxiliary point sources, $\bar{\bar{G}}_\nu \left(\bar{\bar{\Lambda}}_l^{\text{T}} \bar{I}_j\right)$ in the last term of (4.23), are computed at all nodes of the uniform grid via the blocked-FFT algorithm explained in Section 6.7. The fields at the present time step are locally interpolated from the auxiliary grid back onto the primary surface mesh by the left matrix multiplication $\bar{\bar{\Lambda}}_l$ in (4.24).

For planar geometries, a planar uniform grid $\sum_{q=1}^{M^2}$ coincide with the original triangular grid is employed [110, 111]. Clearly, the AIM is well-suited for large relatively flat surfaces. As an alternative approach, the Green's function, instead of the basis functions, can be interpolated to the uniform Cartesian grid on the rectangular bounding box to decouple the source and receivers [112]. It can also be used in fast Gaussian gridding framework [113].

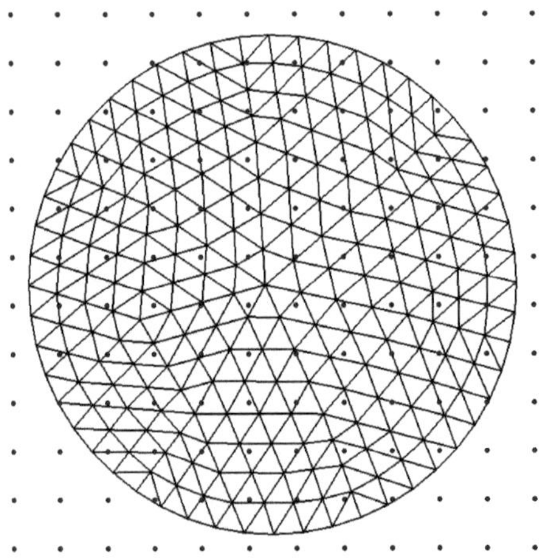

Figure 4.7: The process of projecting the original edge sources on triangular meshes to the AIM uniform grid.

Chapter 5

Near–Field Computations

Once the transient current on the scatterer are found, the calculation of the resulting radiated electric and magnetic fields is a straightforward process [1] as far as the observation point is not located on the surface and most of the functions from the current evaluation stage can be reused. This widely admitted statement has caused direct field calculation routines to be overlooked as a simple post-processing stages in the integral equation-related literature. In many practical problems as those opposed in Section 6.11, accurate near-field calculation is of great importance. Yet there are few well-documented algorithms specifically aimed the direct near-field computations. In [1], in order to calculate the near electric and magnetic fields at an arbitrarily point $\mathbf{r}(x, y, z)$, the curl and gradient operators are approximated by finite difference formulations involving the evaluation of the field quantities, namely the vector and scaler potentials, in *six* neighboring points $\mathbf{r}(x \pm \Delta x, y \pm \Delta y, z \pm \Delta z)$. This is computationally unacceptable. Besides, a time derivative is taken from (2.2) which in turn imposes an extra numerical integration over time at the final stage of near electric field calculations in [1]. In a relatively similar context, the moment methods themselves, however, invoke the calculation of potential integrals due to the source subdomains at close observation points. This chapter intends to introduce general approaches for direct field calculations exactly at the desired point, without using any approximation for curl and gradient operating on scalar and vector potentials. Unlike the earlier integration methods designed to decouple the spatial and temporal integrations by numerical approximations, analytical approaches are reviewed in this chapter to integrate in space-time concurrently. Eventually, each one of the following sections lead to more efficient and accurate potential calculations which can also be incorporated into the Galerkin's schemes in the former chapters.

5.1 Closed-Form Fields of Linear Potentials

The discretized TD EFIE is converted to a matrix equation system by Galerkin testing in space and point matching in time [5]. The surface integral over the observer patches is approximated by the value of the respective integrand at the centroid of the subdomain, $\rho_m^{\pm} \rightarrow \rho_m^{c\pm}$. To calculate the elements of the discretization matrices, instead of redundant direct evaluation of the mutual coupling of edge pairs $\mathbf{f}_m$ individually, subdomain pairs S_q interactions are considered. In comparison with the direct edge-by-edge combinations, this matrix fill-in approach increases up to 9, 12, and 16 times the computational efficiency in evaluating the mutual impedance of the edges belonging to triangle-triangle, triangle-rectangle, and rectangle-rectangle pairs, respectively. Wilton compact formulas for the potentials due to linearly varying source distributions on polygonal patch cells [48] avoid numerical integrations and significantly facilitate evaluating the interaction of triangle or quadrilateral source subdomains on the testing subdomain regardless of their shapes. Fig. 5.1 demonstrates all the quantities appearing in the analytical closed-form of interior space integrals in the MoM can be obtained by projecting the observation point onto the plane containing the source subdomain [5, 101, 114, 115].

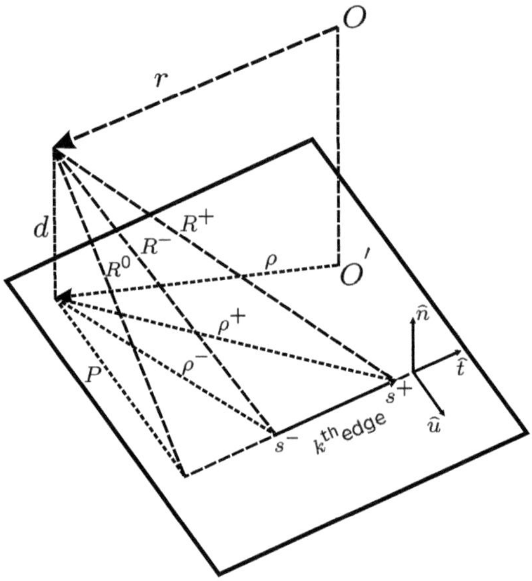

Figure 5.1: All the parameters in (5.1) need to be defined for Wilton analytical formulas for potentials due to linearly varying source distribution on polygonal patches.

To formulate the analytical calculation of all inner double integrals (with the assumption of constant time signature for patches within time intervals) in Section 2.5, the following two well-known contour integrals are used [116]

$$I_i^{-1} = \int_{\partial_i T_k^q} \frac{1}{|\mathbf{r}-\mathbf{r}'|} \mathrm{d}l = \ln\left(\frac{R_i^+ + s_i^+}{R_i^- + s_i^-}\right)$$

$$I_i^1 = \int_{\partial_i T_k^q} |\mathbf{r}-\mathbf{r}'| \mathrm{d}l = \frac{1}{2}\left(s_i^+ R_i^+ - s_i^- R_i^- + R_i^{0^2} I_i^{-1}\right)$$

where

$$\begin{aligned}
\hat{u}_i &= \hat{t}_i \times \hat{n} \\
d &= (\mathbf{r} - \mathbf{r}_i^{\pm}) \cdot \hat{n} \\
s_i^{\pm} &= (\mathbf{r}_i^{\pm} - \mathbf{r}) \cdot \hat{t}_i \\
P_i &= (\mathbf{r}_i^{\pm} - \mathbf{r}) \cdot \hat{u}_i \\
R_i^0 &= \sqrt{P_i^2 + d^2} \\
R_i^{\pm} &= \sqrt{R_i^{0\,2} + s_i^{\pm 2}}.
\end{aligned} \tag{5.1}$$

The closed-form expressions for the surface integrals (2.46)-(2.49) on the source subdomains (linear potentials) can be found in [117]. Defining d as the distance of the observation point from the plane the triangular source patch laid in

$$I_{-3} = \int_{T_k^q} \frac{1}{|\mathbf{r} - \mathbf{r}'|^3} \mathrm{d}\mathbf{r}' = \frac{1}{|d|} \sum_{i=1}^{3} [\arctan\left(\frac{P_i s_i^+}{R_i^{0\,2} + dR_i^+}\right) - \arctan\left(\frac{P_i s_i^-}{R_i^{0\,2} + dR_i^-}\right)] \tag{5.2}$$

and $I_{-3} = 0$ when $d = 0$. When the polynomial temporal basis function are applied, the eventual combination of $\frac{T(t-\frac{R}{c})}{c\partial t}$ times (2.48) plus $T(t - \frac{R}{c})$ times (2.49) cancels out the potential term $\frac{1}{R^2}$, and hence no $\int_{T_k^q} \frac{1}{|\mathbf{r}-\mathbf{r}'|^2}\mathrm{d}\mathbf{r}'$ is required anymore and (5.2) alone suffices for the MFIE matrix construction. For the analytical evaluation of the magnetic vector potential and the electric scalar potential in the EFIE, one may use [48]

$$I_{-1} = \int_{T_k^q} \frac{1}{|\mathbf{r} - \mathbf{r}'|} \mathrm{d}\mathbf{r}' = \sum_{i=1}^{3} P_i I_i^{-1} - d^2 I_{-3} \tag{5.3}$$

$$\mathbf{I}_{-1_i} = \int_{T_k^q} \frac{\mathbf{r}' - \mathbf{r}_i}{|\mathbf{r} - \mathbf{r}'|} \mathrm{d}\mathbf{r}' = \sum_{i=1}^{3} \hat{\mathbf{u}}_i I_i^1 + (\mathbf{r} - d\hat{\mathbf{n}} - \mathbf{r}_i) I_{-1} \tag{5.4}$$

where $\partial_i T_k^q$, $i = 1, 2, 3$ are the edges of the source triangle T_k^q and when the source patch is quadrilateral $\partial_i S_k^q$, $i = 1, 2, 3, 4$. These closed-form expressions are traditionally used only for singularity subtraction (extraction) in the frequency-domain MoM. In the TDBE, they can be utilized to evaluate all matrix elements. The recent three formulas can be alternatively used for the evaluation of (5.8), $k = -3, -1$ and (5.7), $k = -1$ on straight segments in the next section. For the MFIE with barycentric retarded time approximation or alternative form of (D)EFIE (2.16), one can also use [116]

$$\mathbf{I}_{-3} = \int_{T_k^q} \frac{\mathbf{r} - \mathbf{r}'}{|\mathbf{r} - \mathbf{r}'|^3} \mathrm{d}\mathbf{r}' = \sum_{i=1}^{3} \hat{\mathbf{u}}_i I_i^{-1} + \hat{\mathbf{n}}\, dI_{-3}. \tag{5.5}$$

When quadratic time evolution is considered, one additionally needs

$$I_1 = \int_{T_k^q} |\mathbf{r} - \mathbf{r}'| \mathrm{d}\mathbf{r}' = \frac{1}{3} \left(\sum_{i=1}^{3} P_i I_i^1 + d^2 I_{-1} \right). \tag{5.6}$$

5.2 Closed-Form Fields of Time-Varying RWG Sources

This section demonstrates that the convolution between transient RWG sources with polynomial signatures and the 3D time domain free space Green's function can be evaluated exactly [118]. Once the assumption of unchanging within time intervals is avoided for the current elements to refrain that the coefficients $I_k(\tau)$ be pulled out of the surface integrals in (2.42), (2.43), and (2.44), one comes up with

$$\frac{\mu}{4\pi}\sum_{k=1}^{M}\int_{S_m}\mathbf{f}_m(\mathbf{r})\cdot\int_{S_k}\frac{\partial T(\tau)}{\partial\tau}\frac{\mathbf{f}_k(\mathbf{r}')}{R}\mathrm{d}S'\mathrm{d}S$$
$$+\frac{1}{4\pi\epsilon}\sum_{k=1}^{M}\int_{S_m}\nabla_{\mathbf{r}}\cdot\mathbf{f}_m(\mathbf{r})\int_{S_k}\left[\int_0^{\tau}T(t')\mathrm{d}t'\right]\frac{\nabla_{\mathbf{r}'}\cdot\mathbf{f}_k(\mathbf{r}')}{R}\mathrm{d}S'\mathrm{d}S=\int_{S_m}\mathbf{f}_m(\mathbf{r})\cdot\underline{\mathbf{E}}^i(\mathbf{r},t_n)\mathrm{d}S$$

$$\frac{\mu}{4\pi}\sum_{k=1}^{M}\int_{S_m}\mathbf{f}_m(\mathbf{r})\cdot\int_{S_k}\frac{\partial^2 T(\tau)}{\partial\tau^2}\frac{\mathbf{f}_k(\mathbf{r}')}{R}\mathrm{d}S'\mathrm{d}S$$
$$+\frac{1}{4\pi\epsilon}\sum_{k=1}^{M}\int_{S_m}\nabla_{\mathbf{r}}\cdot\mathbf{f}_m(\mathbf{r})\int_{S_k}\nabla_{\mathbf{r}'}\cdot\mathbf{f}_k(\mathbf{r}')\frac{T(\tau)}{R}\mathrm{d}S'\mathrm{d}S=\int_{S_m}\mathbf{f}_m(\mathbf{r})\cdot\frac{\partial\underline{\mathbf{E}}^i(\mathbf{r},t_n)}{\partial t}\mathrm{d}S$$

$$\frac{1}{2}\sum_{k=1}^{M}T(t_n)\int_{S_m}\mathbf{f}_m(\mathbf{r})\cdot\mathbf{f}_k(\mathbf{r}')\mathrm{d}S-\frac{1}{4\pi}\left[\sum_{k=1}^{M}\int_{S_m}\mathbf{f}_m(\mathbf{r})\cdot\hat{\mathbf{n}}\times\int_{S_k}\frac{\partial T(\tau)}{c\partial t}\frac{\mathbf{f}_k(\mathbf{r}')\times\mathbf{R}}{R^2}\mathrm{d}S'\mathrm{d}S\right.$$
$$\left.+\sum_{k=1}^{M}\int_{S_m}\mathbf{f}_m(\mathbf{r})\cdot\hat{\mathbf{n}}\times\int_{S_k}T(\tau)\frac{\mathbf{f}_k(\mathbf{r}')\times\mathbf{R}}{R^3}\mathrm{d}S'\mathrm{d}S\right]=\int_{S_m}\mathbf{f}_m(\mathbf{r})\cdot\hat{\mathbf{n}}\times\mathbf{H}^i(\mathbf{r},t_n)\mathrm{d}S.$$

where $\tau = n\Delta t - \frac{R}{c}$. The temporal basis function $T(t)$ is piecewise continuous, with different expression when its argument is within regions $[-\Delta t, 0], [0, \Delta t], \ldots, [(p-1)\Delta t, p\Delta t]$. As a result, integrands in (5.7)-(5.7) are piecewise continuous over the domain of integration S_k [119]. The domain of integration thus can be partitioned into a set of subregions in which the function is continuous $S_k = \sum_{\beta=1} S_\beta$, as illustrated in Fig. 5.2. Fictitious concentric spheres emanating from arbitrary observation point $\mathbf{r}$ and dilating with radii $c\Delta t$, $2c\Delta t, \ldots$ intersect the plane containing the source triangle S_k in a series of circles, Fig. 5.2. Depending on the time step size (oversampling factor α), some of the circles cut through the triangular source patch (Fig. 5.3).

The time bases $T(\tau)|_{\tau=n\Delta t-\frac{R}{c}}$ in (3.36)-(3.39), (3.43), and (3.44) can be expressed as a polynomial of order p with respect to its argument [120]

$$T(t-|\mathbf{r}-\mathbf{r}'|/c)=a_0+a_1(t-|\mathbf{r}-\mathbf{r}'|/c)+a_2(t-|\mathbf{r}-\mathbf{r}'|/c)^2+\ldots+a_p(t-|\mathbf{r}-\mathbf{r}'|/c)^p.$$

This permits exact evaluation of source integrals. For instance, when $p = 2$, the second order Lagrange (3.37), for the MFIE

$$\int_{S_k}\left[\frac{\partial_t T(t-|\mathbf{r}-\mathbf{r}'|/c)}{c|\mathbf{r}-\mathbf{r}'|^2}+\frac{T(t-|\mathbf{r}-\mathbf{r}'|/c)}{|\mathbf{r}-\mathbf{r}'|^3}\right]\mathrm{d}\mathbf{r}'$$
$$=(a_0+a_1t+a_2t^2)\int_{S_k}\frac{1}{|\mathbf{r}-\mathbf{r}'|^3}\mathrm{d}\mathbf{r}'-\frac{a_2}{c^2}\int_{S_k}\frac{1}{|\mathbf{r}-\mathbf{r}'|}\mathrm{d}\mathbf{r}'$$

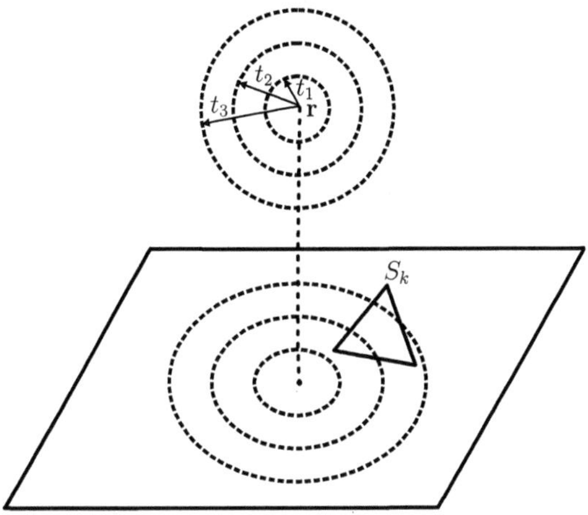

Figure 5.2: The time-growing sphere originated from the observation point intersects with source triangular subdomains and the integration range is subdivided.

and for the EFIE

$$\int_{S_k} \frac{T(t-|\mathbf{r}-\mathbf{r}'|/c)}{|\mathbf{r}-\mathbf{r}'|}\mathrm{d}\mathbf{r}' = -\frac{1}{c}(a_1+2a_2t)$$
$$+(a_0+a_1t+a_2t^2)\int_{S_k}\frac{1}{|\mathbf{r}-\mathbf{r}'|}\mathrm{d}\mathbf{r}' + \frac{a_2}{c^2}\int_{S_k}|\mathbf{r}-\mathbf{r}'|\mathrm{d}\mathbf{r}'$$

$$\int_{S_k}\rho\frac{\partial_t^2 T(t-|\mathbf{r}-\mathbf{r}'|/c)}{|\mathbf{r}-\mathbf{r}'|}\mathrm{d}\mathbf{r}' = 2a_2\int_{S_k}\frac{\rho}{|\mathbf{r}-\mathbf{r}'|}\mathrm{d}\mathbf{r}',$$

and when $p=1$, the first order Lagrange (3.36), for the MFIE

$$\int_{S_k}\left[\frac{\partial_t T(t-|\mathbf{r}-\mathbf{r}'|/c)}{c|\mathbf{r}-\mathbf{r}'|^2}+\frac{T(t-|\mathbf{r}-\mathbf{r}'|/c)}{|\mathbf{r}-\mathbf{r}'|^3}\right]\mathrm{d}\mathbf{r}' = (a_0+a_1t)\int_{S_k}\frac{1}{|\mathbf{r}-\mathbf{r}'|^3}\mathrm{d}\mathbf{r}'$$

and for the DEFIE

$$\int_{S_k}\frac{\int_{-\infty}^{t-|\mathbf{r}-\mathbf{r}'|/c}T(t')\mathrm{d}t'}{|\mathbf{r}-\mathbf{r}'|}\mathrm{d}\mathbf{r}' = -\frac{1}{c}(a_0+2a_1t)$$
$$+(a_1+a_0t+a_1\frac{t^2}{2})\int_{S_k}\frac{1}{|\mathbf{r}-\mathbf{r}'|}\mathrm{d}\mathbf{r}' + \frac{t}{c^2}(\frac{a_1}{2c^2})\int_{S_k}|\mathbf{r}-\mathbf{r}'|\mathrm{d}\mathbf{r}'$$

$$\int_{S_k}\rho\frac{\partial_t T(t-|\mathbf{r}-\mathbf{r}'|/c)}{|\mathbf{r}-\mathbf{r}'|}\mathrm{d}\mathbf{r}' = a_1\int_{S_k}\frac{\rho}{|\mathbf{r}-\mathbf{r}'|}\mathrm{d}\mathbf{r}'.$$

Therefore, all the time-varying linear potential integrals entail linear combinations of the following integrals

$$\mathbf{I}_k = \int_{S_\beta}\rho|\mathbf{r}-\mathbf{r}'|^k\mathrm{d}\mathbf{r}', \quad k=-3 \text{ and } -1,0,1,\ldots \tag{5.7}$$

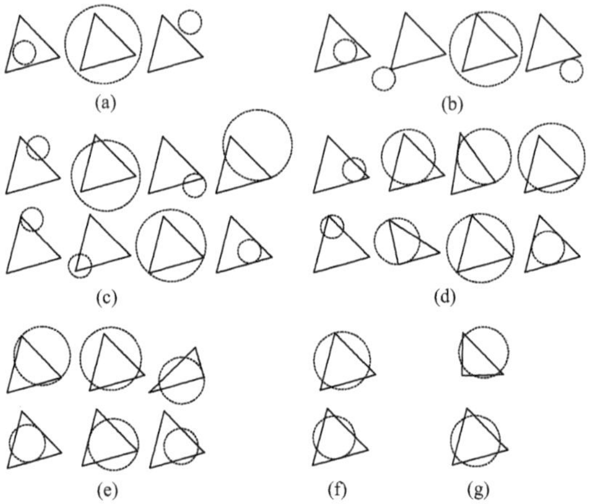

Figure 5.3: All possible geometrical relationships between a triangular subdomain and a sphere. (a)-(g) illustrate different sorted cases with respectively 0 to 6 intersection points: (a) no intersection and (b) 1 intersection point, (c) 2, (d) 3, (e) 4, (f) 5, and (g) 6 intersection points.

$$I_k = \int_{S_\beta} |\mathbf{r} - \mathbf{r}'|^k \mathrm{d}\mathbf{r}', \quad k = -3 \text{ and } -1, 0, 1, \dots. \tag{5.8}$$

Using the Gauss theorem in two dimensions, the surface integrals in (5.7) and (5.8) are transformed into contour line integrals along the boundary of the subregions ∂S_β,

$$\mathbf{I}_k = \frac{1}{k+2} \oint_{\partial S_\beta} \hat{\mathbf{d}} |\mathbf{r} - \mathbf{r}'|^{k+2} \mathrm{d}l, \quad k = -3 \text{ and } -1, 0, 1, \dots$$

$$\begin{aligned} I_k &= \frac{1}{k+2} \int_{S_\beta} \nabla \cdot \left(\frac{|\mathbf{r} - \mathbf{r}'|^{k+2}}{\rho} \hat{\rho} \right) \mathrm{d}\mathbf{r}' \\ &= \frac{1}{k+2} \oint_{\partial S_\beta} \frac{|\mathbf{r} - \mathbf{r}'|^{k+2}}{\rho} \hat{\rho} \cdot \hat{\mathbf{d}} \mathrm{d}l - \frac{1}{k+2} \int_{\partial(\epsilon \bigcap S_\beta)} \frac{|\mathbf{r} - \mathbf{r}'|^{k+2}}{\rho} \mathrm{d}l, \quad k = -3 \text{ and } -1, 0, 1, \dots \end{aligned}$$

where $\hat{\mathbf{d}}$ is the outward normal to ∂S_β and ϵ is a circle in the plane defined by S_k centered at $\mathbf{r}_p$ and with radius approaching zero. The vector $\mathbf{r}_p$ is the projection of $\mathbf{r}$ on the plane in which the source triangle is laid, the vector $\vec{\rho}$ begins from $\mathbf{r}_p$ and points to arbitrarily points on the subregions, $\hat{\rho} = \frac{\vec{\rho}}{\rho}$, $\rho = |\vec{\rho}|$.

The boundary ∂S_{kq} is generally composed of straight segments and arcs, where the straight segments $\overline{\mathbf{r}_1'\mathbf{r}_2'}$ are part of the boundary of S_β, and the arcs $\widehat{\mathbf{r}_1'\mathbf{r}_2'}$ arise from the intersection between S_β and fictitious spheres [119]. When integrating over a straight segment $\overline{\mathbf{r}_1'\mathbf{r}_2'}$, starting with point $\mathbf{r}_1'$ and ending at point $\mathbf{r}_2'$, a local Cartesian coordinate system $(\hat{\mathbf{u}}, \hat{\mathbf{v}}, \hat{\mathbf{n}})$ with the origin $\mathbf{r}_p$ is established. Axis n is aligned with the normal to the plane S_β laid in and axes u and v are defined as

$$\hat{\mathbf{u}} = \frac{\mathbf{r}_2' - \mathbf{r}_1'}{|\mathbf{r}_2' - \mathbf{r}_1'|} \qquad \hat{\mathbf{v}} = \hat{\mathbf{n}} \times \hat{\mathbf{u}}. \tag{5.9}$$

When the integration variable $\mathbf{r}'$ sweeps along the straight segment, vector $\hat{\mathbf{d}} = \pm\hat{\mathbf{v}}$ has a constant direction and $\hat{\rho} \times \hat{\mathbf{d}} = \pm v_0/\rho$ where $v_0 = (\mathbf{r}_p - \mathbf{r}'_1)\cdot\hat{\mathbf{v}}$. Therefore, the line integrals are parameterized by

$$\frac{1}{k+2}\int_{\mathbf{r}'_1\mathbf{r}'_2} \hat{\mathbf{d}}|\mathbf{r}-\mathbf{r}'|^{k+2} \mathrm{d}l = \frac{\hat{\mathbf{d}}}{k+2}\int_{u_1}^{u_2} \mathrm{d}u \left[\sqrt{u^2+v_0^2+n_0^2}\right]^{k+2}$$

$$\frac{1}{k+2}\int_{\mathbf{r}'_1\mathbf{r}'_2} \frac{|\mathbf{r}-\mathbf{r}'|^{k+2}}{\rho}\hat{\rho}\cdot\hat{\mathbf{d}}\mathrm{d}l = \frac{\pm v_0}{k+2}\int_{u_1}^{u_2} \mathrm{d}u\frac{\left[\sqrt{u^2+v_0^2+n_0^2}\right]^{k+2}}{u^2+v_0^2}$$

where $u_1 = (\mathbf{r}'_1 - \mathbf{r}_p)\cdot\hat{\mathbf{u}}$, $u_2 = (\mathbf{r}'_2 - \mathbf{r}_p)\cdot\hat{\mathbf{u}}$, and $n_0 = |\mathbf{r} - \mathbf{r}_p| = d_{\text{in Section }5.1}$. Alternative formulation can be found in [116].

When integrating over an arc segment $\widehat{\mathbf{r}'_1\mathbf{r}'_2}$, starting with $\mathbf{r}'_1$ and ending at $\mathbf{r}'_2$, $\hat{\mathbf{d}} = \pm\hat{\rho}$ and $\rho = \rho_0 = |\mathbf{r}'_1 - \mathbf{r}_p| = |\mathbf{r}'_2 - \mathbf{r}_p|$ are constant.

$$\frac{1}{k+2}\int_{\widehat{\mathbf{r}'_1\mathbf{r}'_2}} \hat{\mathbf{d}}|\mathbf{r}-\mathbf{r}'|^{k+2} \mathrm{d}l = \frac{2\sin(\phi/2)}{k+2}(\mathbf{r}'_3 - \mathbf{r}_p)\left[\sqrt{\rho_0^2+n_0^2}\right]^{k+2}$$

$$\frac{1}{k+2}\int_{\widehat{\mathbf{r}'_1\mathbf{r}'_2}} \frac{|\mathbf{r}-\mathbf{r}'|^{k+2}}{\rho}\hat{\rho}\cdot\hat{\mathbf{d}}\,\mathrm{d}l = \frac{\pm\phi}{k+2}\left[\sqrt{\rho_0^2+n_0^2}\right]^{k+2}$$

where $\mathbf{r}'_3$ is the mid-point on the arc, and ϕ is the angle spanned by the arc with respect to $\mathbf{r}_p$. Specifically, if $\mathbf{r}_p$ is completely within S_β, then $\partial(\epsilon \bigcap S_\beta)$ is a complete circle and $\phi = 2\pi$; if $\mathbf{r}_p$ is on a straight segment of ∂S_β, then $\phi = \pi$; and when $\mathbf{r}_p$ is at the intersection of two straight segments of ∂S_β, ϕ is their angle of intersection. It is apparent that $\mathbf{r}_p$ never falls on an arc segment of ∂S_β.

5.2.1 Precise Evaluation of the MOT Four-Fold Integrals

Analytical evaluation of three out of the four space integrals for every MOT matrix elements has been presented recently in [121]. To facilitate the evaluation of the quadric spatial integrations, a local Cartesian coordinate system (u, v, n) is constructed at $\mathbf{r}_i, i = 1, 2, 3$

$$\begin{aligned}
\hat{\mathbf{n}} &= \frac{(\mathbf{r}_2 - \mathbf{r}_1)\times(\mathbf{r}_3 - \mathbf{r}_1)}{|(\mathbf{r}_2 - \mathbf{r}_1)\times(\mathbf{r}_3 - \mathbf{r}_1)|} \\
\hat{\mathbf{n}}' &= \frac{(\mathbf{r}'_2 - \mathbf{r}'_1)\times(\mathbf{r}'_3 - \mathbf{r}'_1)}{|(\mathbf{r}'_2 - \mathbf{r}'_1)\times(\mathbf{r}'_3 - \mathbf{r}'_1)|} \\
\hat{\mathbf{u}} &= \hat{\mathbf{n}}\times\hat{\mathbf{n}}' \\
\hat{\mathbf{v}} &= \hat{\mathbf{n}}\times\hat{\mathbf{u}}
\end{aligned}$$

where $(\mathbf{r}_1, \mathbf{r}_2, \mathbf{r}_3)$ are vertices of the observation triangle S_m with unit normal vector $\hat{\mathbf{n}}$ and $(\mathbf{r}'_1, \mathbf{r}'_2, \mathbf{r}'_3)$ are vertices of the source triangle S_k with unit normal vector $\hat{\mathbf{n}}'$. The integration over the observation S_m and source S_k subdomains can be transformed to the whole space $\mathbf{r} = u\hat{\mathbf{u}} + v\hat{\mathbf{v}} + n\hat{\mathbf{n}}$, $\mathrm{d}\mathbf{r} = \mathrm{d}u\mathrm{d}v\mathrm{d}n$

$$\begin{aligned}
&\int_{S_m} \nabla_{\mathbf{r}} \cdot \mathbf{f}_m(\mathbf{r}) \int_{S_k} \nabla_{\mathbf{r}'} \cdot \mathbf{f}_k(\mathbf{r}') \frac{T(\tau)}{R} \mathrm{d}S' \mathrm{d}S = \iint_{S_m} \iint_{S_k} \frac{T(\tau)}{R} \mathrm{d}\mathbf{r}' \mathrm{d}\mathbf{r} \\
&= \iiint \prod_m(\mathbf{r}) \iiint \prod_k(\mathbf{r}') \frac{T(\tau)}{R} \mathrm{d}\mathbf{r}' \mathrm{d}\mathbf{r} = \iiint \prod_m(\mathbf{r}) \iiint \prod_k(\mathbf{r} - \mathbf{R}) \frac{T(\tau)}{R} \mathrm{d}\mathbf{r}' \mathrm{d}\mathbf{r} \\
&= \iiint \underbrace{\iiint \prod_m(\mathbf{r}) \prod_k(\mathbf{r} - \mathbf{R}) \mathrm{d}\mathbf{r}}_{\Omega(\mathbf{R})} \frac{T(\tau)}{R} \mathrm{d}\mathbf{r}'
\end{aligned} \tag{5.10}$$

by using the Jacobian $\prod(u, v, n) = \bar{\prod}(u, v)\delta(n)$ where $\delta(.)$ denotes the Dirac delta function and $\bar{\prod}(u, v) = 1$ if (u, v) reside in S and 0 elsewhere. The spatial correlation function Ω has a compact support, i.e. a finite three-dimensional volume. [1] To calculate the field due to a volume source observed at $\mathbf{R} = u''\hat{\mathbf{u}} + v''\hat{\mathbf{v}} + n''\hat{\mathbf{n}}$ in (5.10), the integral is decomposed along n''

$$\begin{aligned}
&\iiint \Omega(\mathbf{R}) \frac{T(\tau)}{R} \mathrm{d}\mathbf{R} = \int_{-\infty}^{\infty} \int_{-\infty}^{\infty} \int_{-\infty}^{\infty} \Omega(\mathbf{R}) \frac{T(\tau)}{R} \mathrm{d}u'' \mathrm{d}v'' \mathrm{d}n'' \\
&= \int_{-\infty}^{\infty} \mathrm{d}\bar{n} \left[\int_{-\infty}^{\infty} \int_{-\infty}^{\infty} \int_{-\infty}^{\infty} [\Omega(u'', v'', \bar{n}) \delta(n'' - \bar{n})] \frac{T(\tau)}{R} \mathrm{d}u'' \mathrm{d}v'' \mathrm{d}n'' \right]
\end{aligned} \tag{5.11}$$

In (5.11), $\Omega(u'', v'', \bar{n})\delta(n'' - \bar{n})$ is the slice of $\Omega(u'', v'', n'')$ at $n'' = \bar{n}$. This slice is the spatial correlation between the observation triangle S_m and the line segment that comprises the intersection between plane $n = n''$ and the source triangle S_k. The correlation between S_m and the line segment yields a polygon. Surface integration over this polygon can be evaluated analytically [121], while the integration over $\bar{n}$ in (5.11) is carried out numerically using Gauss-Legendre rule.

5.3 Polar Integration for Space-Time Quadratures

Considering the time-varying current on the surface of the doublet source subdomains $S_k = S_k^+ + S_k^-$ ($S_k = \sum_q S_k^q$, $q = \pm$) associated with the edge k in (2.37)

$$\mathbf{J}_k(\mathbf{r}', t) = \begin{cases} \frac{\mu}{4\pi} I_k \sum_q \frac{l_k}{2A_k^q} \rho'^q_k T(t), & \mathbf{r}' \in S_k^q \\ 0, & \mathbf{r}' \notin S_k^q. \end{cases} \tag{5.12}$$

Although S_k^q is primarily assumed to be triangular patch T_k^q, it can be quadrilateral P_k^q as Section 2.4.2 or any dual combination of linearly-varying flat multilateral facets, e.g. $S_k = T_k^+ + P_k^-$. The magnetic vector potential at the field point $\mathbf{r}$ due to the current density $\mathbf{J}_k(\mathbf{r}', t)$ (2.4)

$$\mathbf{A}_k(\mathbf{r}, t) = \frac{\mu}{4\pi} I_k \sum_q \frac{l_k}{2A_k^q} \int_{S_k^q} \frac{\rho'^q_k T(t - \frac{|\mathbf{r} - \mathbf{r}'|}{c})}{|\mathbf{r} - \mathbf{r}'|} \mathrm{d}S'$$

[1] When S_m and S_k are co-planar, the support of Ω lies in a surface.

$$\frac{\partial^2}{\partial t^2}\mathbf{A}_k(\mathbf{r},t) = \frac{\mu}{4\pi} I_k \sum_q \frac{l_k}{2A_k^q} \int_{S_k^q} \frac{\rho_k'^q \ddot{T}(t - \frac{|\mathbf{r}-\mathbf{r}'|}{c})}{|\mathbf{r}-\mathbf{r}'|} \mathrm{d}S'$$

and the electric scalar Hertz potential due to the charge density $\sigma_k(\mathbf{r}', t)$ (2.12)

$$\Phi_k(\mathbf{r},t) = -\frac{1}{4\pi\epsilon} I_k \sum_q \frac{l_k}{A_k^q} \int_{S_k^q} \frac{T(t - \frac{|\mathbf{r}-\mathbf{r}'|}{c})}{|\mathbf{r}-\mathbf{r}'|} \mathrm{d}S'.$$

In case of magnetic field formulations, one encounters the following two terms for the curl of magnetic vector potential (2.23) (also for the dual case, electric vector potential in (2.28))

$$\frac{1}{\mu}\nabla \times \mathbf{A}_k(\mathbf{r},t) = \frac{1}{c}\frac{\partial}{\partial t}\mathbf{A}_k^1(\mathbf{r},t) + \mathbf{A}_k^2(\mathbf{r},t) \tag{5.13}$$

where

$$\frac{\partial}{\partial t}\mathbf{A}_k^1(\mathbf{r},t) = \frac{1}{4\pi} I_k \sum_q \frac{l_k}{2A_k^q} \int_{S_k^q} \frac{\rho_k'^q \times (\mathbf{r}-\mathbf{r}')\dot{T}(t - \frac{|\mathbf{r}-\mathbf{r}'|}{c})}{|\mathbf{r}-\mathbf{r}'|^2} \mathrm{d}S'.$$

$$\mathbf{A}_k^2(\mathbf{r},t) = \frac{1}{4\pi} I_k \sum_q \frac{l_k}{2A_k^q} \int_{S_k^q} \frac{\rho_k'^q \times (\mathbf{r}-\mathbf{r}')T(t - \frac{|\mathbf{r}-\mathbf{r}'|}{c})}{|\mathbf{r}-\mathbf{r}'|^3} \mathrm{d}S'.$$

Note that the singularity is removed in the latter cases through Cauchy PV ($S \to S_0$) in (2.19).

The polar quadrature transform the surface integration in 3D space into a 2D integration in the plane of the source subdomain [25] by

$$\rho_k^q = \rho - \rho^{kq} \tag{5.14}$$

$$R = \sqrt{\rho^2 + d^2} \tag{5.15}$$

$$\mathrm{d}S' = \rho\,\mathrm{d}\rho\,\mathrm{d}\phi \tag{5.16}$$

where d is the distance from the observation point to its projection on the plane of the subdomain, ρ is the vector from the projection point to the source point, ρ^{kq} is the vector from the projection point to the free vertex of the q side of the subdomain k and is constant with respect to the integration. When the transformations (5.14)-(5.16) are applied and the local vector $\rho = |\rho|(\hat{\mathbf{u}}\cos\theta + \hat{\mathbf{v}}\sin\theta)$ is decomposed into its local Cartesian components,

$$\begin{aligned}
\hat{\mathbf{n}} &= \frac{(\mathbf{r}_2 - \mathbf{r}_1) \times (\mathbf{r}_3 - \mathbf{r}_1)}{|(\mathbf{r}_2 - \mathbf{r}_1) \times (\mathbf{r}_3 - \mathbf{r}_1)|} \\
\hat{\mathbf{u}} &= \frac{\mathbf{r}_2 - \mathbf{r}_1}{|\mathbf{r}_2 - \mathbf{r}_1|} \\
\hat{\mathbf{v}} &= \hat{\mathbf{u}} \times \hat{\mathbf{n}}
\end{aligned}$$

where $(\mathbf{r}_1, \mathbf{r}_2, \mathbf{r}_3)$ are vertices of the source triangle S_k lying on the plane $(\hat{\mathbf{u}}, \hat{\mathbf{v}})$ with unit normal vector $\hat{\mathbf{n}}$, one obtains

$$\frac{\partial^2}{\partial t^2}\mathbf{A}_k(\mathbf{r},t) = \frac{\mu}{4\pi} I_k \sum_q \frac{l_k}{2A_k^q} [\ddot{\mathbf{P}} - \rho^{kq}\ddot{\mathrm{P}}]$$

$$\Phi_k(\mathbf{r},t) = -\frac{1}{4\pi\epsilon} I_k \sum_q \frac{l_k}{A_k^q} \mathrm{P}.$$

where

$$\mathbf{P} = \int\int \frac{\rho^2(\hat{\mathbf{u}}\cos\theta + \hat{\mathbf{v}}\sin\theta)T(t - \frac{\sqrt{\rho^2+d^2}}{c})}{\sqrt{\rho^2+d^2}} \mathrm{d}\rho\, \mathrm{d}\theta \tag{5.17}$$

$$\mathrm{P} = \int\int \frac{\rho T(t - \frac{\sqrt{\rho^2+d^2}}{c})}{\sqrt{\rho^2+d^2}} \mathrm{d}\rho\, \mathrm{d}\theta. \tag{5.18}$$

For the MFIE $\rho_k'^q \times (\mathbf{r}-\mathbf{r}') = (\rho - \rho^{kq}) \times (d\hat{\mathbf{n}} - \rho) = d(\rho - \rho^{kq}) + \rho \times \rho^{kq}$, in which the last term is terminated, since later on $\hat{\mathbf{n}} \times \left(\frac{1}{\mu}\nabla \times \mathbf{A}_k(\mathbf{r},t)\right)$ causes $\hat{\mathbf{n}} \times \rho \times \rho^{kq} = 0$. Therefore,

$$\frac{\partial}{\partial t}\mathbf{A}_k^1(\mathbf{r},t) = \frac{\mathrm{d}}{4\pi} I_k \sum_q \frac{l_k}{2A_k^q}[\dot{\mathbf{P}}^1 - \rho^{kq}\dot{\mathrm{P}}^1]$$

$$\mathbf{A}_k^2(\mathbf{r},t) = \frac{\mathrm{d}}{4\pi} I_k \sum_q \frac{l_k}{2A_k^q}[\mathbf{P}^2 - \rho^{kq}\mathrm{P}^2]$$

where

$$\mathbf{P}^1 = \int\int \frac{\rho^2(\hat{\mathbf{u}}\cos\theta + \hat{\mathbf{v}}\sin\theta)T(t - \frac{\sqrt{\rho^2+d^2}}{c})}{\rho^2+d^2} \mathrm{d}\rho\, \mathrm{d}\theta, \quad \mathrm{P}^1 = \int\int \frac{\rho T(t - \frac{\sqrt{\rho^2+d^2}}{c})}{\rho^2+d^2} \mathrm{d}\rho\, \mathrm{d}\theta \tag{5.19}$$

$$\mathbf{P}^2 = \int\int \frac{\rho^2(\hat{\mathbf{u}}\cos\theta + \hat{\mathbf{v}}\sin\theta)T(t - \frac{\sqrt{\rho^2+d^2}}{c})}{(\rho^2+d^2)^{\frac{3}{2}}} \mathrm{d}\rho\, \mathrm{d}\theta, \quad \mathrm{P}^2 = \int\int \frac{\rho T(t - \frac{\sqrt{\rho^2+d^2}}{c})}{(\rho^2+d^2)^{\frac{3}{2}}} \mathrm{d}\rho\, \mathrm{d}\theta \tag{5.20}$$

The temporal derivatives are left for analytical evaluation through $T(t)$ in (3.36)-(3.39), (3.43), and (3.44). The above integrals are separated into functions of the two variables (ρ,θ), e.g.,

$$\mathrm{P} = \int_\rho \int_\theta \varrho(\rho)\vartheta(\theta)\, \mathrm{d}\rho\, \mathrm{d}\theta$$

where

$$\varrho(\rho) = \frac{\rho T(t - \frac{\sqrt{\rho^2+d^2}}{c})}{\sqrt{\rho^2+d^2}} \tag{5.21}$$

$$\vartheta(\theta) = \begin{cases} \cos\theta, & \hat{\mathbf{u}} \text{ component } (5.17) \\ \sin\theta, & \hat{\mathbf{v}} \text{ component } (5.17) \\ 1, & (5.18). \end{cases} \tag{5.22}$$

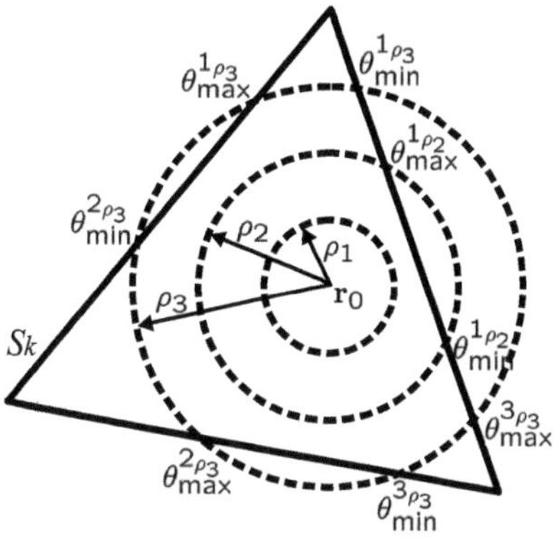

Figure 5.4: The circle-triangle intersections (the solid arcs ending points) determine θ integration limits when the radius ρ grows with time discretely. The origin $\mathbf{r}_0$ is the projection of observation point on the plane S_k.

Apparently, the transformation removes the singularity in the EFIE, i.e., when $d = 0$ (as in the singular case)

$$\varrho(\rho) = T(t - \frac{\rho}{c})$$

which is non-singular. In the MFIE, the contribution of the singularity at $R = 0$ has been already removed in S_0 (2.19).

The limits of the integration in θ depend on the value of ρ. As illustrated in Fig. 5.4, the circle defined by ρ may intersect a triangle between zero and six places, creating between one and three arc intervals to analytically integrate in θ. If there are no intersection points, the interval is $(0,2\pi)$.

$$\begin{aligned} \int_{\theta(\rho)} \vartheta(\theta)\mathrm{d}\theta = \begin{pmatrix} \varrho_c(\rho) \\ \varrho_s(\rho) \\ \varrho_1(\rho) \end{pmatrix} &= \sum_{i=1}^{K(\rho)} \int_{\theta^i_{\min}(\rho)}^{\theta^i_{\max}(\rho)} \begin{pmatrix} \cos\theta \\ \sin\theta \\ 1 \end{pmatrix} \mathrm{d}\theta \\ &= \sum_{i=1}^{K(\rho)} \begin{pmatrix} \sin\theta^i_{\max}(\rho) - \sin\theta^i_{\min}(\rho) \\ -\cos\theta^i_{\max}(\rho) + \cos\theta^i_{\min}(\rho) \\ \theta^i_{\max}(\rho) - \theta^i_{\min}(\rho) \end{pmatrix} \end{aligned} \tag{5.23}$$

where $K(\rho)$ is 1, 2, or 3 depending on ρ. Therefore, the original 2D integrations on $(\hat{\mathbf{u}}, \hat{\mathbf{v}})$ plane are reduced to a 1D integral along ρ,

$$\mathbf{P} = \int_{\rho_{\min}}^{\rho_{\max}} \rho\varrho(\rho)[\hat{\mathbf{u}}\varrho_c(\rho) + \hat{\mathbf{v}}\varrho_s(\rho)]\mathrm{d}\rho \tag{5.24}$$

$$\mathrm{P} = \int_{\rho_{\min}}^{\rho_{\max}} \varrho(\rho)\varrho_1(\rho)\mathrm{d}\rho. \tag{5.25}$$

The integration in ρ can be computed with a 1D Newton-Cotes or Gauss-Legendre quadrature rule [64].

It should be pointed out that the integration limits in (5.24) and (5.25), $[\rho_{\min}, \rho_{\max}]$ and $[\theta^i_{\min}(\rho), \theta^i_{\max}(\rho)]$, are written somewhat naively to keep the derivation simple. Only for certain values of ρ, the collection of points in S_k that are a distance ρ away from the origin will form a single arc that lies in the range $[\theta^i_{\min}, \theta^i_{\max}]$. These arcs are formed by the intersection of the triangle S_k with a sphere of radius $R = ct$ (or equivalently, with the circle whose center $\mathbf{r}_0$ is the projection of observation point on the plane S_k and radius is $\rho(t) = \sqrt{(ct)^2 - d^2}$, (5.15). To determine $[\theta^i_{\min}, \theta^i_{\max}]$, first, the intersection(s) of the circle with the line segments that bound the triangle are found. Then, these intersection points are grouped in pairs, each of which defines an arc segment within S_k.

The intersection points of the circle with triangle sides (line segments) are obtained in the two-dimensional local (Cartesian) coordinate system (u, v). Hence, the S_k vertices $\mathbf{r}_i = (x_i, y_i, z_i)$ are expressed in (u, v) coordiantes, i.e. $\mathbf{r}_i = (u_i, v_i, n = 0)$ where $u_i = (\mathbf{r}_i - \mathbf{r}_0)\cdot\hat{\mathbf{u}}$ and $v_i = (\mathbf{r}_i - \mathbf{r}_0)\cdot\hat{\mathbf{v}}$. Knowing the line segment is bounded by $\mathbf{r}_i$ and $\mathbf{r}_j$ vertices where $(i \neq j)$, and defining

$$\begin{aligned} d_u &= u_j - u_i && (5.26)\\ d_v &= v_j - v_i && (5.27)\\ d_r &= \sqrt{d_u^2 + d_v^2} && (5.28)\\ D &= \begin{vmatrix} u_i & u_j \\ v_i & v_j \end{vmatrix} = u_i v_j - v_i u_j && (5.29)\\ \Delta &= \sqrt{\rho^2(t) d_r^2 - D^2} \end{aligned}$$

gives coordinates of intersection points

$$\begin{aligned} u_{1,2} &= \frac{D d_v \pm \operatorname{sgn}(d_v) d_u \Delta}{d_r^2} && (5.30)\\ v_{1,2} &= \frac{-D d_u \pm |d_v| \Delta}{d_r^2} \end{aligned}$$

where

$$\operatorname{sgn}(x) := \begin{cases} -1, & x < 0 \\ 1, & \text{otherwise} \end{cases}.$$

The number of line-circle intersection points would be

$$\begin{cases} 0, & \Delta < 0 \\ 1, & \Delta = 0 \\ 2, & \Delta > 0 \end{cases}$$

and only those intersection points out of (u_1, v_1) and (u_2, v_2) are acceptable that fall within the line segment $\overline{\mathbf{r}_i\mathbf{r}_j}$ (not on its extension). To identify the pairs of intersection points that define arcs that completely lie within S_k, the intersection points are sorted in increasing order of the associated azimuthal angle $0 \leq \phi < 2\pi$ values, measured from the unit vector $\hat{\mathbf{u}}$.

$$\phi = \tan^{-1}\left(\frac{v_{1,2}}{u_{1,2}}\right).$$

A duplicate of the first intersection point $\phi_1 + 2\pi$ is added to the end of sorted list of points to account the last arc. The midpoint of every two consecutive intersection points (the arc segments) are checked whether they coincide within the interior of S_k or not. The midpoint of the arc segment can be found in local coordinates by multiplying the radius $\rho(t)$ with the arc unit bisector vector

$$\hat{\mathbf{e}}(t) = \hat{\mathbf{u}}\cos(\frac{\phi_i + \phi_{i+1}}{2}) + \hat{\mathbf{v}}\sin(\frac{\phi_i + \phi_{i+1}}{2}). \tag{5.31}$$

To check whether the midpoint of the arc $\rho(t)(\mathrm{e}_u(t), \mathrm{e}_v(t), 0)$ lies in the interior of S_k, the point is converted to the barycentric (area) coordinates $(\lambda_1, \lambda_2, \lambda_3 = 1 - \lambda_1 - \lambda_2)$,

$$\begin{pmatrix} \lambda_1 \\ \lambda_2 \end{pmatrix} = \begin{pmatrix} u_1 - u_3 & u_2 - u_3 \\ v_1 - v_3 & v_2 - v_3 \end{pmatrix}^{-1} \begin{pmatrix} \rho \mathrm{e}_u - u_3 \\ \rho \mathrm{e}_v - v_3 \end{pmatrix}.$$

If $0 < \lambda_i < 1,\ i = 1, 2, 3$, the mid-point of the arc is located inside S_k and the arc contributes to $\sum_{i=1}^{K(\rho)}$ in (5.24) and (5.25). One may alternatively check whether the point and free vertex are on the same side of the triangle edges using the source code in Appendix 8.7. When the circle does not intersect with any edges of the triangle, i.e., $\rho(t)$ is located outside of S_k, the arc contribution is automatically zero. This case corresponds to the time intervals for which the fields of the current distribution over S_k have either not arrived at the observer location or have already passed it. The latter situation also applies to the case when the observation point is located inside S_k and $\rho(t)$ is longer than the distance between and the furthest vertex to this point. No singularity also arises when the observation point falls within S_k and $\rho(t)$ is shorter than the distance between and the nearest edge to the observer, since the entire circle resides in S_k and thus

$$\begin{pmatrix} \varrho_c(\rho) \\ \varrho_s(\rho) \\ \varrho_1(\rho) \end{pmatrix} = \begin{pmatrix} 0 \\ 0 \\ 2\pi \end{pmatrix}.$$

5.3.1 A Nyström Method without Local Corrections

Unlike the Galerkin method, which discretizes the integral equations with testing the field effects of a set of basis functions on the same set of associated subdomains, the Nyström method solves for samples of the current at the nodes of an integration rule [122, 123]. As in the Galerkin method, the scatterer is assumed to be composed of M patches, $S = \sum_{k=1}^{M} S_k$. On each patch an integration rule of the form

$$\int_{S_k} f(\rho)\mathrm{d}\rho = \sum_i \omega_i f(\rho_i)$$

is performed, where ρ_i are the position vectors at the integration nodes on patch S_k and ω_i are the integration weights for the Gauss-Legendre integration rule. The discretization is accomplished by sampling the incident field and the current at the testing nodes $\mathbf{r}_m$ similar to the point-matching method, so (2.9) becomes

$$\sum_{k=1}^{M} \left[\frac{\partial^2 \mathbf{A}_k(\mathbf{r}_m, t)}{\partial t^2} + \nabla \Phi_k(\mathbf{r}_m, t) \right]_{\tan} = \frac{\partial \underline{\mathbf{E}}^i(\mathbf{r}_m, t)}{\partial t} \qquad m = 1, 2, \dots M$$

where $\mathbf{A}_k$ and Φ_k have already been evaluated in this section;

$$\frac{\mu}{4\pi} \sum_{k=1}^{M} \sum_q \frac{l_k}{2A_k^q} \left([\ddot{\mathbf{P}} - \rho^{kq}\ddot{\mathrm{P}}] - 2c^2 \mathrm{P} \right) I_k = \frac{\partial \underline{\mathbf{E}}^i(\mathbf{r}_m, t_n)}{\partial t} \qquad m = 1, 2, \dots M.$$

5.3.2 Analytical Evaluation of Arc Length and Bisecting Vector

A more efficient technique to evaluate the arc lengths without calculating the intersection points of the sphere $R = ct$ with the edges of the subdomains is presented in the following lines. Let C_i represent the ith edge of subdomain S_k^q with length l_i and λ_i denote the barycentric coordinates of the projection point $\mathbf{r}_0$ on S_k^q. For evaluation of the integrand ϱ_1 in (5.23), one may alternatively consider

$$\varrho_1(\zeta) = \sum_{i=1}^{K(\rho)} \operatorname{sgn}(\lambda_i)\alpha_i(\zeta) \tag{5.32}$$

where α_i is the arc length measured from the radial extension of C_i edge(s) corner to the circle center $\mathbf{r_0}$ and should not be confused with θ^i in previous section. To determine α_i, let $l_{i,\min}$ and $l_{i,\max}$ respectively denote the smaller and larger distances to the ends of the line segment C_i from the point $\mathbf{r}_0$ with the angles $\phi_{i,\min}$ and $\phi_{i,\max}$ with respect to the axis $\mathbf{u}$. Also, let a_i represent the vertical distance from $\mathbf{r}_0$ to the (extension of) the line segment C_i. When the vertical distance lies out of the line segment C_i, $l_{i,\max}^2 > l_{i,\min}^2 + l_i^2$ as seen in Fig. 5.5(a)

$$\alpha_i(\zeta) = \begin{cases} \cos^{-1}(\frac{a_i}{l_{i,\max}}) - \cos^{-1}(\frac{a_i}{l_{i,\min}}) & \zeta < l_{i,\min} \\ \cos^{-1}(\frac{a_i}{l_{i,\max}}) - \cos^{-1}(\frac{a_i}{\zeta}) & l_{i,\min} < \zeta < l_{i,\max} \\ 0 & l_{i,\max} < \zeta. \end{cases} \tag{5.33}$$

When the vertical distance falls on the line segment C_i, i.e., $l_{i,\max}^2 \leq l_{i,\min}^2 + l_i^2$, two arc emerge as illustrated in Fig. 5.5(b) and

$$\alpha_i(\zeta) = \alpha_{i,\min}(\zeta) + \alpha_{i,\max}(\zeta) \tag{5.34}$$

$$\alpha_{i,\mathrm{m}}(\zeta) = \begin{cases} \cos^{-1}(\frac{a_i}{l_{i,\mathrm{m}}}) & \zeta < a_i \\ \cos^{-1}(\frac{a_i}{l_{i,\mathrm{m}}}) - \cos^{-1}(\frac{a_i}{\zeta}) & a_i \leq \zeta < l_{i,\mathrm{m}} \\ 0 & l_{i,\mathrm{m}} < \zeta \end{cases} \tag{5.35}$$

where $\mathrm{m} \in \{\max, \min\}$.

The bisecting vector $\mathbf{e} = \rho[\hat{\mathbf{u}}\varrho_c(\rho) + \hat{\mathbf{v}}\varrho_s(\rho)]$ in (5.23), can also be evaluated considering the sign of barycentric coordinates λ_i,

$$e_u(\zeta) = \zeta \sum_{i=1}^{K(\zeta)} \operatorname{sgn}(\lambda_i) e_{u,i}(\zeta) \tag{5.36}$$

$$e_v(\zeta) = \zeta \sum_{i=1}^{K(\zeta)} \operatorname{sgn}(\lambda_i) e_{v,i}(\zeta) \tag{5.37}$$

where $e_{u,i}(\zeta)$ and $e_{v,i}(\zeta)$ are projections of the intersection of the growing sphere and the ith edge of the triangle to the $\hat{\mathbf{u}}$ and $\hat{\mathbf{v}}$ axes. Again, there are again two cases to consider. When $l_{i,\max}^2 > l_{i,\min}^2 + l_i^2$ as in Fig. 5.5(a)

$$e_{u,i}(\zeta) = \chi \begin{cases} \sin[\phi_{i,\max}] - \sin[\phi_{i,\min}] & \zeta < l_{i,\min} \\ \sin[\phi_{i,\max}] - \sin[\phi_{i,\max} - \chi\alpha_i(\zeta)] & l_{i,\min} < \zeta < l_{i,\max} \\ 0 & l_{i,\max} < \zeta \end{cases} \tag{5.38}$$

$$e_{v,i}(\zeta) = \chi \begin{cases} \cos[\phi_{i,\max}] - \cos[\phi_{i,\min}] & \zeta < l_{i,\min} \\ \cos[\phi_{i,\max}] - \cos[\phi_{i,\max} - \chi\alpha_i(\zeta)] & l_{i,\min} < \zeta < l_{i,\max} \\ 0 & l_{i,\max} < \zeta \end{cases} \tag{5.39}$$

and when $l_{i,\max}^2 \leq l_{i,\min}^2 + l_i^2$, as in Fig. 5.5(b)

$$e_{u,i}(\zeta) = \chi \begin{cases} \sin[\phi_{i,\max}] - \sin[\phi_{i,\min}] & \zeta < a_i \\ \sin[\phi_{i,\max}] - \sin[\phi_{i,\max} - \chi\alpha_{i,\max}(\zeta)] & \\ -\sin[\phi_{i,\min}] + \sin[\phi_{i,\min} + \chi\alpha_{i,\min}(\zeta)] & a_i < \zeta < l_{i,\min} \\ \sin[\phi_{i,\max}] - \sin[\phi_{i,\max} - \chi\alpha_{i,\max}(\zeta)] & l_{i,\min} < \zeta < l_{i,\max} \\ 0 & l_{i,\max} < \zeta \end{cases}$$

$$e_{v,i}(\zeta) = \chi \begin{cases} \cos[\phi_{i,\max}] - \cos[\phi_{i,\min}] & \zeta < a_i \\ \cos[\phi_{i,\max}] - \cos[\phi_{i,\max} - \chi\alpha_{i,\max}(\zeta)] & \\ -\cos[\phi_{i,\min}] + \cos[\phi_{i,\min} + \chi\alpha_{i,\min}(\zeta)] & a_i < \zeta < l_{i,\min} \\ \cos[\phi_{i,\max}] - \cos[\phi_{i,\max} - \chi\alpha_{i,\max}(\zeta)] & l_{i,\min} < \zeta < l_{i,\max} \\ 0 & l_{i,\max} < \zeta \end{cases}$$

where

$$\chi = \begin{cases} -1 & \phi_{i,\max} - \phi_{i,\min} > \pi \\ 1 & \phi_{i,\max} - \phi_{i,\min} < \pi \end{cases} \tag{5.40}$$

assuming $\phi_{i,\max} - \phi_{i,\min}$ is in $[0, 2\pi)$. Obviously, when $R < d$, the sphere and the triangle surface do not intersect, and thus, $\alpha(\zeta) = 0$ and $\mathbf{e}(\zeta) = 0$.

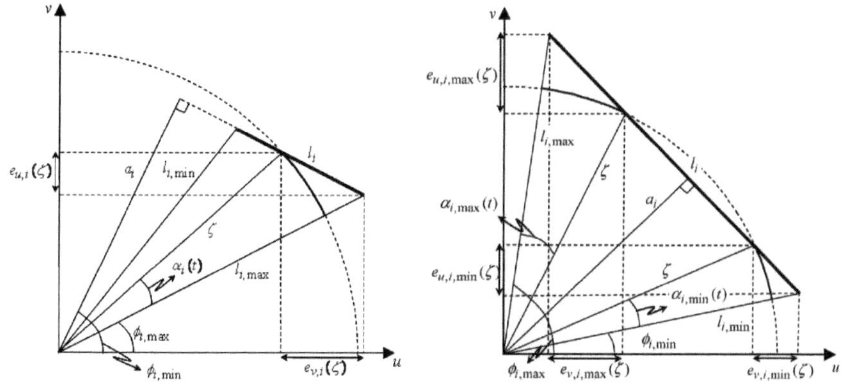

Figure 5.5: The required parameters to calculate the arc length and the bisecting vector [124].

5.4 Exact Evaluation of Retarded Potential Integrals

The vector potential expression in (2.4),

$$\mathbf{A}(\mathbf{r},t) = \frac{\mu}{4\pi}\int_S \frac{\mathbf{J}(\mathbf{r}',t-\frac{R}{c})}{R}\mathrm{d}S',$$

is originally

$$\mathbf{A}(\mathbf{r},t) = \frac{\mu}{4\pi}\int_S \mathbf{J}(\mathbf{r}',t) * \frac{\delta(t-\frac{R}{c})}{R}\mathrm{d}S'$$

where $*$ denotes temporal convolution. Discretizing the current density on the radiating body S in terms of spatial basis functions as in (2.33) gives

$$\mathbf{A}(\mathbf{r},t) = \frac{\mu}{4\pi}\sum_{k=1}^{M} I_k(t) * \int_S \mathbf{f}_k(\mathbf{r}')\frac{\delta(t-\frac{R}{c})}{R}\mathrm{d}S'.$$

Using the RWG functions, the integral becomes

$$\mathbf{A}(\mathbf{r},t) = \frac{\mu}{4\pi}\sum_{k=1}^{M} \frac{ql_k}{2A_k^q} I_k(t) * \underbrace{\int_{S_k} (\mathbf{r}'-\mathbf{r}_k^q)\frac{\delta(t-\frac{R}{c})}{R}\mathrm{d}S'}_{\mathbf{H}_k(\mathbf{r},t)}.$$

Let ρ, ρ', and ρ_k^q respectively represent the projection of $\mathbf{r}$, $\mathbf{r}'$, and $\mathbf{r}_k^q$ on the plane that includes S_k. Then, $\mathbf{r}'-\mathbf{r}_k^q = (\rho'-\rho)+(\rho-\rho_k^q)$ gives

$$\mathbf{H}_k(\mathbf{r},t) = \underbrace{\int_{S_k} (\rho'-\rho)\frac{\delta(t-\frac{R}{c})}{R}\mathrm{d}S'}_{\mathbf{H}'_k(\mathbf{r},t)} + (\rho-\rho_k^q)\underbrace{\int_{S_k} \frac{\delta(t-\frac{R}{c})}{R}\mathrm{d}S'}_{H_k(\mathbf{r},t)}.$$

Also, let (ζ,ϕ,n) denote the cylindrical coordinates of the local Cartesian system $(\hat{\mathbf{u}},\hat{\mathbf{v}},\hat{\mathbf{n}})$ defined in Section 5.3. The scalar integral is written in the new local coordinate system as

$$\begin{aligned} H_k(\mathbf{r},t) &= \int_{\zeta_{\min}}^{\zeta_{\max}}\int_{\theta_{\min}(\zeta)}^{\theta_{\max}(\zeta)} \frac{\delta(t-\frac{R}{c})}{R}\zeta\mathrm{d}\phi\mathrm{d}\zeta \\ &= \int_{\zeta_{\min}}^{\zeta_{\max}} \alpha(\zeta)\frac{\delta(t-\frac{R}{c})}{R}\zeta\mathrm{d}\zeta \end{aligned} \tag{5.41}$$

where $\alpha(\zeta)=\sum_{i=1}^{K(\zeta)}\theta_{\max}^i(\zeta)-\theta_{\min}^i(\zeta)$, identical to $\varrho_1(\rho)$ in Section 5.3, is the sum of the length of the arcs (in radians) formed by the intersection of a sphere centered at $\mathbf{r}$ and the radius $R=ct$ and the triangle S_k. To evaluate the outer integral, the integration variable is changed from ζ to $R=\sqrt{\zeta^2+d^2}$

$$\begin{aligned} H_k(\mathbf{r},t) &= \int_{R_{\min}}^{R_{\max}} \alpha(\sqrt{R^2-d^2})\delta(t-\frac{R}{c})dR \\ &= \begin{cases} \alpha(\sqrt{(ct)^2-d^2}) & R_{\min}<ct<R_{\max} \\ 0 & \text{elsewhere} \end{cases} \end{aligned} \tag{5.42}$$

Repeating the procedure for the scalar potential also yields the demand for (5.42),

$$\begin{aligned}\phi(\mathbf{r},t) &= \frac{1}{4\pi\epsilon}\sum_{k=1}^{M} I_k(t) * \int_S \nabla_{\mathbf{r}'}\cdot\mathbf{f}_k(\mathbf{r}')\frac{\delta(t-\frac{R}{c})}{R}\mathrm{d}S' \\ &= \frac{1}{4\pi\epsilon}\sum_{k=1}^{M}\frac{ql_k}{A_k^q}I_k(t) * \underbrace{\int_{S_k}\frac{\delta(t-\frac{R}{c})}{R}\mathrm{d}S'}_{H_k(\mathbf{r},t)}.\end{aligned}$$

The vector integral can also be written in the local cylindrical coordinates centered at ρ as

$$\begin{aligned}\mathbf{H}'_k(\mathbf{r},t) &= \int_{S_k}(\rho'-\rho)\frac{\delta(t-\frac{R}{c})}{R}\mathrm{d}S' \\ &= \int_{\zeta_{\min}}^{\zeta_{\max}}\int_{\theta_{\min}(\zeta)}^{\theta_{\max}(\zeta)}(\zeta\cos\phi\hat{\mathbf{u}}+\zeta\sin\phi\hat{\mathbf{v}})\frac{\delta(t-\frac{R}{c})}{R}\zeta\mathrm{d}\phi\mathrm{d}\zeta \\ &= \int_{\zeta_{\min}}^{\zeta_{\max}}[\varrho_c(\zeta)\hat{\mathbf{u}}+\varrho_s(\zeta)\hat{\mathbf{v}}]\frac{\delta(t-\frac{R}{c})}{R}\zeta^2\mathrm{d}\zeta \\ &= \int_{R_{\min}}^{R_{\max}}[\varrho_c(\sqrt{R^2-d^2})\hat{\mathbf{u}}+\varrho_s(\sqrt{R^2-d^2})\hat{\mathbf{v}}]\delta(t-\frac{R}{c})\sqrt{R^2-d^2}dR \\ &= [\varrho_c(\sqrt{(ct)^2-d^2})\hat{\mathbf{u}}+\varrho_s(\sqrt{(ct)^2-d^2})\hat{\mathbf{v}}]\times \\ \frac{1}{2}&\left[ct\sqrt{(ct)^2-d^2}-d^2\ln\left(ct+\sqrt{(ct)^2-d^2}\right)\right] \quad R_{\min}<ct<R_{\max} \quad (5.43)\end{aligned}$$

where the evaluation of ϱ_c and ϱ_s functions has been already explained in Section 5.3.

5.5 Far-Field Approximations and RCS Calculations

The scattered magnetic field is related to the induced transient current by (2.3)

$$\mathbf{H}^s(\mathbf{r},t) = \frac{1}{4\pi}\nabla_{\mathbf{r}}\times\int_S\frac{\mathbf{J}(\mathbf{r}',\tau)}{R}\mathrm{d}S'.$$

Taking the curl operator inside the integral and using the vector identity (2.21) results in

$$\mathbf{H}^s(\mathbf{r},t) = \frac{1}{4\pi}\int_S\frac{\nabla\times\mathbf{J}(\mathbf{r}',\tau)}{R}-\frac{\mathbf{R}}{R^3}\times\mathbf{J}(\mathbf{r}',\tau)\,\mathrm{d}S'.$$

The far-scattered fields representing the radar signature of the objects are defined in the observation distances $R^2 >> R$ where R is the distance from the center of the object [1, 27]. Therefore, one can neglect the second term of the integrand. Using (2.22)

$$\nabla\times\mathbf{J}(\mathbf{r}',\tau) = \frac{1}{c}\frac{\partial}{\partial t}\mathbf{J}(\mathbf{r}',\tau)\times\frac{\hat{\mathbf{R}}}{R}$$

where $\hat{\mathbf{R}}$ is a unit vector in direction $\mathbf{R}$. The far field approximation also lets directions $\hat{\mathbf{R}}\approx\hat{\mathbf{r}}$, amplitude $R\approx r$ and the retarded time $\tau\approx t-\frac{r-\mathbf{r}'\cdot\hat{\mathbf{r}}}{c}$ where $\hat{\mathbf{r}}=\frac{\mathbf{r}}{r}$ and $r=|\mathbf{r}|$. Thus, substitution of current expansion (3.35) gives

$$\mathbf{H}^s(\mathbf{r},t)\approx\frac{1}{4\pi c}\sum_{k=1}^{M}I_k\sum_q\int_{S_k^q}\frac{\partial T(\tau)}{\partial t}\frac{\mathbf{f}_k(\mathbf{r}')\times\hat{\mathbf{r}}}{r}\mathrm{d}S'.$$

Using the definition of basis functions, e.g., (2.37), the surface integral can be carried out analytically

$$\sum_q \int_{S_k^q} \frac{\partial T(\tau)}{\partial t} \frac{\mathbf{f}_k(\mathbf{r}') \times \hat{\mathbf{r}}}{r} \mathrm{d}S' = \frac{l_k}{2r} \sum_q \frac{\partial T(\tau)}{\partial t} \rho_k^{cq} \times \hat{\mathbf{r}}.$$

So, the normalized far magnetic field is obtained by

$$r\mathbf{H}^s(\mathbf{r}, t_n) \approx -\frac{1}{8\pi c} \hat{\mathbf{r}} \times \sum_{k=1}^{M} l_k I_k^n \sum_q \rho_k^{cq} \frac{\partial T(\tau_{k,n}^q)}{\partial t} \tag{5.44}$$

where $\tau_{k,n}^q = t_n - \frac{(r - \mathbf{r}_k^{cq} \cdot \hat{\mathbf{r}})}{c}$. Consequently, the far-scattered electric field is given by

$$\mathbf{E}^s(\mathbf{r}, t_n) = \eta \mathbf{H}^s(\mathbf{r}, t_n) \times \hat{\mathbf{r}} \tag{5.45}$$

and the radar cross section (RCS) is found through

$$\sigma = \lim_{r \to \infty} 4\pi r^2 \frac{|\mathbf{E}^s|^2}{|\mathbf{E}^i|^2}. \tag{5.46}$$

Alternatively, one may start the far field computations directly by calculating the scattered electric field

$$\mathbf{E}^s(\mathbf{r}, t) \approx -\frac{\partial}{\partial t} \mathbf{A}(\mathbf{r}, t).$$

From (2.4) and (3.35), it gives

$$\mathbf{E}^s(\mathbf{r}, t) \approx -\frac{\mu}{4\pi} \sum_{k=1}^{M} I_k \sum_q \int_{S_k^q} \frac{\partial T(\tau)}{\partial t} \frac{\mathbf{f}_k(\mathbf{r}')}{r} \mathrm{d}S'$$

which ends up with a similar expression when one uses first (5.44) and then (5.45)

$$r\mathbf{E}^s(\mathbf{r}, t_n) \approx -\frac{\mu}{8\pi} \sum_{k=1}^{M} l_k I_k^n \sum_q \rho_k^{cq} \frac{\partial T(\tau_{k,n}^q)}{\partial t}. \tag{5.47}$$

When the (A)MOD methods have been already applied for the current evaluation, the term $I_k^n \frac{\partial T(\tau_{k,n}^q)}{\partial t}$ in (5.47) has to be substituted by (3.48)

$$\frac{\mathrm{d}}{\mathrm{d}t} c_k(\tau_{k,n}^q) = s \sum_{j=0}^{N} \left(\frac{1}{2} c_{k,j} + \sum_{\iota=0}^{j-1} c_{k,\iota} \right) \phi_j(s\tau_{k,n}^q). \tag{5.48}$$

or (3.91)

$$\frac{\mathrm{d}}{\mathrm{d}t} c_k(\tau_{k,n}^q) = -s \sum_{j=0}^{N} c_{k,j} \left(\frac{1}{2} \phi_j(s\tau_{k,n}^q) + \sum_{\iota=0}^{j-1} \phi_\iota(s\tau_{k,n}^q) \right). \tag{5.49}$$

Chapter 6

Numerical Results and Discussion

In all of the examples which follow, the incident electric field is assumed to be a Gaussian-shaped pulse

$$\mathbf{E}^i(\mathbf{r},t) = \mathbf{E}_0 \frac{4}{\sqrt{\pi}T} e^{-\sigma^2} \qquad \sigma = \frac{4}{T}(ct - ct_0 - \mathbf{r}\cdot\hat{\mathbf{k}}) \tag{6.1}$$

where $\hat{\mathbf{k}}$ is the unit vector in the propagation direction of the incident wave, T is the Gaussian pulse width (implies that when $ct - ct_0 - \mathbf{r}\cdot\hat{\mathbf{k}} = \pm\frac{T}{2}$, the exponential decays to approximately 2 percent of its peak value). The parameter t_0 denotes the time instant at which the pulse peak reaches the origin of the reference frame. In the following, we assume that the incident electric field propagates along the z-axis direction $\hat{\mathbf{k}} = -\hat{\mathbf{z}}$ and is polarized along the x-axis with $\mathbf{E}_0 = 120\pi\hat{\mathbf{x}}\,(\mathrm{V/m})$ so as to fulfill the condition $\mathbf{E}_0 \cdot \hat{\mathbf{k}} = 0$. Here, $t_0 = 6\,\mathrm{lm}$ and $T = 4\,\mathrm{lm}$, i.e., $ct_0 = 6\,\mathrm{m}$ and $cT = 4\,\mathrm{m}$. Note that one light meter (lm) is the unit of time taken by the EM wave to travel 1.0 meter distance in a homogeneous media. The Gaussian pulse can be modulated to shift the spectrum of the signal to a center frequency $f_0 \neq 0$,

$$\mathbf{E}^i(\mathbf{r},t) = \mathbf{E}_0 \frac{4}{\sqrt{\pi}T} e^{-\sigma^2} \cos\left(\frac{2\pi f_0}{c}(ct - ct_0 - \mathbf{r}\cdot\hat{\mathbf{k}})\right). \tag{6.2}$$

The TDIEs are solved for perfect electric conducting (PEC) objects using the introduced techniques. The time step size is chosen according to a fraction of minimum distance between the centroids of triangles in lm, i.e., $\Delta t = \alpha\frac{R_{\min}}{c}$. The scattering response of the examined three-dimensional structures is here compared with the results of a well established time-domain code. Intentionally, the results are compared with another time-domain code, in spite of most other works [1, 3, 10, 24, 66] that consider frequency-domain counterparts. Here, the obtained induced surface current densities are compared with the result obtained by a converged, high-resolution FIT simulation which is considered as a reference solution. The FIT is a volume discretization method [39] accessible within the commercially available CST Microwave Studio® software package [56].

6.1 Convergence Study

A perfectly conducting cube centered about the origin, 0.2m on a side, is subdivided monotonically into 2, 4, and 8 divisions along the $\hat{\mathbf{x}}$, $\hat{\mathbf{y}}$, and $\hat{\mathbf{z}}$ directions resulting in 48, 192 and 768 triangular patches, respectively, with M = 72, 288, and 1152 common edges.

Choosing the adaptive quadrature error $\varepsilon = 0.1 \times 10^{-6}$, (3.16) is constructed using the theta time integration method according to (3.23) with $\theta = 0.5$ and solved for the time step factor $\alpha = 7.5$. Fig. 6.1 exhibit typical sparsity pattern of $\bar{\bar{Z}}_0$ for the cube with $M = 72$ for two different time steps. In Fig. 6.1(a), the time step size is $\Delta t = 0.0625$lm. Once one solves the problem for four times shorter time step with sparse impedance matrix depicted in Fig. 6.1(b), late-time instability appearers. Solving for four times larger time step $\Delta t = 0.25$lm renders a totally populated coefficient matrix. To make sure from the late-time stability for the current at any points on the structure, the response associated with all edges are probed as they are observed all together in Fig. 6.2.

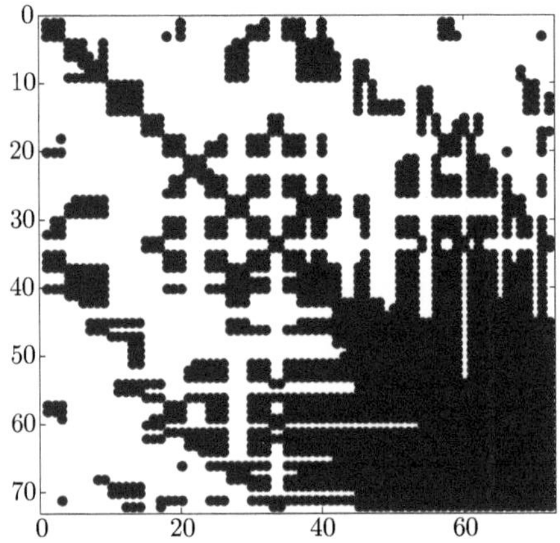

(a) $c\Delta t = 6.25$cm

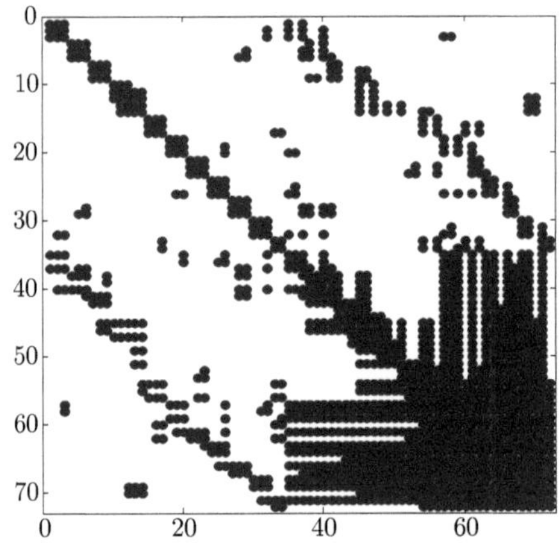

(b) $c\Delta t = 1.5625$cm

Figure 6.1: Typical sparse structure of the coefficient matrix $\bar{\bar{Z}}_0$ in the MOT methods.

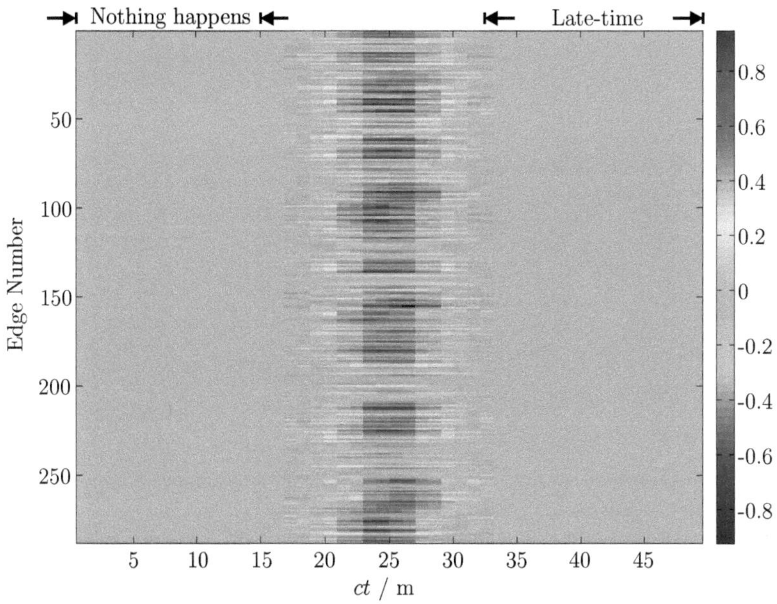

Figure 6.2: Monitoring the late-time current stability at the middle of all triangles' edges of the conducting cube ($M = 288$) when illuminated by the Gaussian plane wave. When $c\Delta t < 15$ the incident field has not yet reached to the object and it leaves the objects for $c\Delta t > 35$, the time interval which is called late time.

The calculated $\hat{\mathbf{x}}$-directed induced current density at the center of the top face of the cube is shown in Fig. 6.3. There is a good agreement between the high resolution FIT results and those obtained using the EFIE method together with the Crank-Nicolson time integration.

Considering the results of employing the converged, high-resolution FIT as the reference solution, Fig. 6.4 demonstrates the relative error, defined by the L^1-norm of the subtraction of the two current vectors, as a function of R_{min} for applying either the implicit backward Euler or Crank-Nicolson integrators. As Fig. 6.4 reveals, in the specified range, one can reach to the 2nd order convergence for the second order Crank-Nicolson time integrator using the first-order time interpolation. Applying the backward Euler method leads to a first-order convergence.

6.2 Consistent Integrator–Interpolator Pairs

The perfect conducting cube (0.2m on a side) is divided monotonically into 2 divisions along the $\hat{\mathbf{x}}, \hat{\mathbf{y}}, \hat{\mathbf{z}}$ directions resulting in 48 triangular patches with 72 common edges. Equations (2.10), (2.11), and (2.24) are solved by the time integrators introduced in Section 3.2 for the unknown induced surface current density $\mathbf{J}(\mathbf{r}, t)$ on the cube at each time step. The magnitude of the x-directed induced surface current in the logarithmic scale at the center of

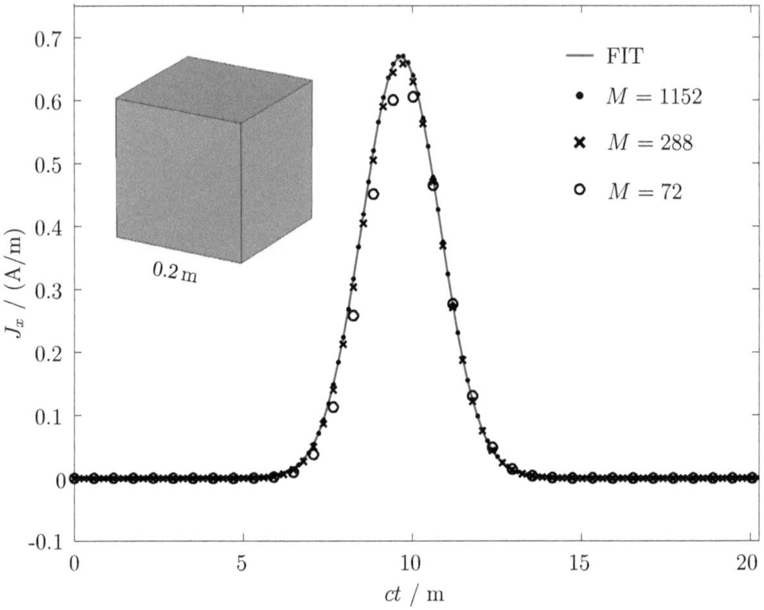

Figure 6.3: Transient current density at the center of the top side of the conducting cube 0.2 m on a side illuminated by the Gaussian plane wave.

the top face of the cube is shown in Fig. 6.5 for the MFIE solved by the implicit (backward) Euler with the time step factor $\alpha = 0.99$, as well as for the EFIE solved by the explicit (forward) Euler with $\alpha = 0.7$, implicit (backward) Euler with $\alpha = 6.66$, the Galerkin method with $\alpha = 5.2$, and the DEFIE solved by the 2nd derivative approximation method with $\alpha = 2.7$. The introduced time integrators in Section 3.2 are all applied in conjunction with the linear interpolation. Any positive slope for the tail of the induced surface current amplitude in logarithmic scale is considered as an index of (late-time) instability of the methods.

As Fig. 6.5 exhibits for the aforementioned values of α the magnitude of the late-time solution of the implicit schemes has no tangible tendency to grow up and thus they are usable for practical purposes. The smaller the values of α, the earlier and mostly the sharper ramps are appeared on the tail of the response in all cases. It is worth noting that in most papers reported stable MOT schemes for numerical solution of the MFIE the backward Euler integrator and linear interpolation have been applied, that is fully consistent with the foundations on DDE behaviour in Section 3.2.3. The adaptive refinement of triangular meshes was used to restrain the relative numerical error in calculation of the potential integrals under the predefined $\epsilon \leq 0.1 \times 10^{-6}$ limit. It is worth noting that reducing the precision of surface quadratures to $\epsilon \leq 10^{-3}$ does not sweep the time instant where the tail of the current starts to rise.

Fig. 6.6 shows the results obtained by solving the EFIE via the second order Crank-Nicolson method and the second order backward difference method (3.29), respectively,

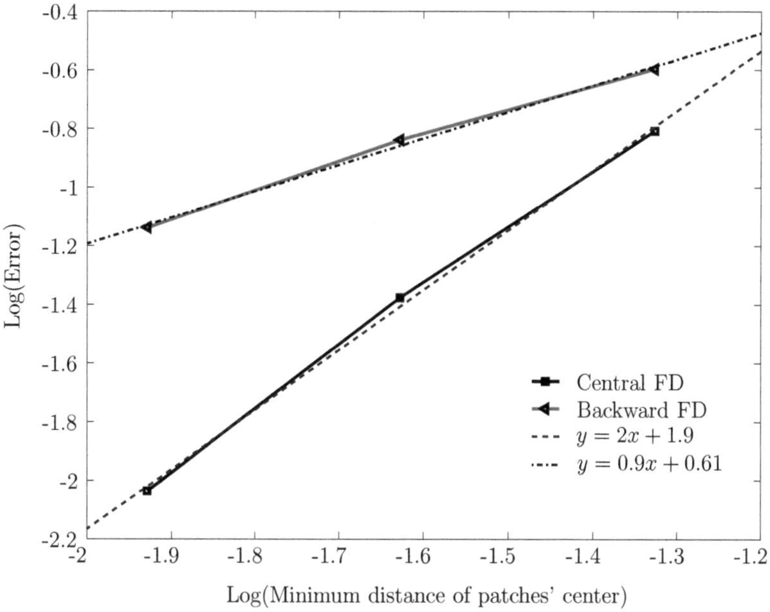

Figure 6.4: Relative L^1-norm error in current evaluation in logarithmic scale with respect to $R^{\min}$, normalized to the cube size.

for $\alpha = 4.7$ and $\alpha = 5.5$. Both integrators cause relatively flat tails in the desired interval when a linear interpolation is used. Basically, one expects the contribution of the integrator in the overall accuracy becomes dominant for small time step sizes whereas the influence of the interpolation outruns over the latter one for large time steps. As conceived by the discussions in the Section 3.2.3, the order of the consistent interpolant for stable results, however, should be one less than that of the integrator. Hence, using quadratic interpolators ($q = 2$) destroys the steadiness of the late-time solution and using even higher order interpolants ($q > 2$) shift up the launch of the linearly growing tail in logarithmic scale to earlier time steps and cause sharper ramps. This set of inappropriate interpolators does not include only the Lagrange polynomials rather as it was observed in the results the use of the cosine square and the cubic spline interpolation also ruins the tranquility of the late-time response.

It is worth mentioning that approximations in evaluation of the outer integrands only at the center of the testing triangles as well as the assumption of no change for the transient current within the individual triangles corrupt the late-time stability of the discussed approaches for smaller values of the α than mentioned. Using symmetric central finite difference to approximate the second derivative, i.e., substituting t_n by t_{n-1} for the last two terms of (3.29)

$$\frac{\mathbf{A}(\mathbf{r}, t_n) - 2\mathbf{A}(\mathbf{r}, t_{n-1}) + \mathbf{A}(\mathbf{r}, t_{n-2})}{\Delta t^2} + \nabla\Phi(\mathbf{r}, t_{n-1}) = \frac{\partial \underline{\mathbf{E}}^i(\mathbf{r}, t_{n-1})}{\partial t},$$

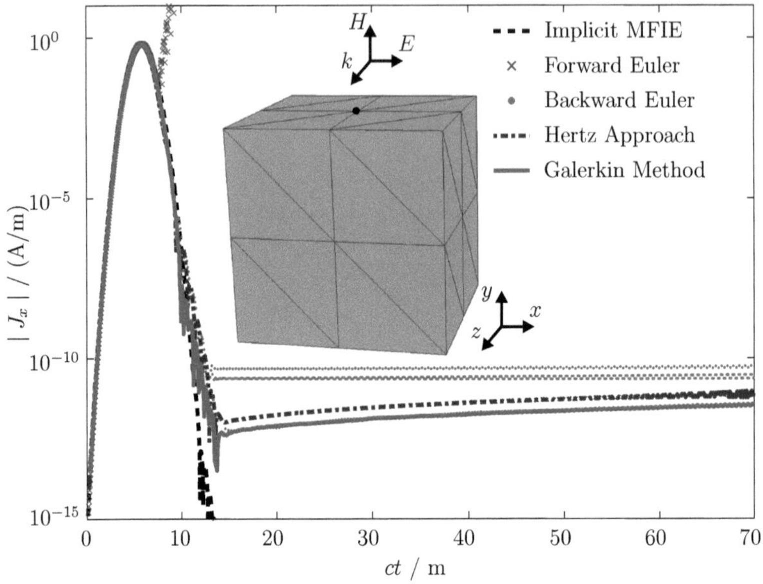

Figure 6.5: Magnitude of the induced surface current at the center of top face of the conducting cube.

gives an approximation of one order higher than (3.29). It ignites a linearly raising tail in the logarithmic scale at very early time stages with any of the discussed interpolation techniques, regardless of the asymmetric version in (3.29). The same statement is true for the most widely known third order backward difference formula

$$\frac{2\mathbf{A}(\mathbf{r},t_n)+3\mathbf{A}(\mathbf{r},t_{n-1})-6\mathbf{A}(\mathbf{r},t_{n-2})+\mathbf{A}(\mathbf{r},t_{n-3})}{6\Delta t}+\nabla\phi(\mathbf{r},t_{n-1})=\underline{\mathbf{E}}^{i}(\mathbf{r},t_{n-1})$$

even with testing of vector potentials not on the barycenter of triangles rather on the middle of the edges, similar to the implementation of explicit regimes [1, 5, 7]. In other words, the use of the two aforementioned difference formula causes totally unstable results, perhaps since even in their implicit version, the scalar potential term does not contribute in construction of the coefficient matrix.

6.3 Subdomain Temporal Basis Functions

In this section, the PEC cube, often used as the benchmark problem for the TDIEs, is posed to demonstrate that using closed-form analytical derivatives provides stability on much smaller time steps than the range approximating derivatives with common finite difference formulas do. To this end, numerical results obtained using the basis functions depicted in Fig. 3.2 are compared with the Crank-Nicolson method with linear Lagrange interpolant, the best (2nd order) choice among the integrator-interpolator couples reported

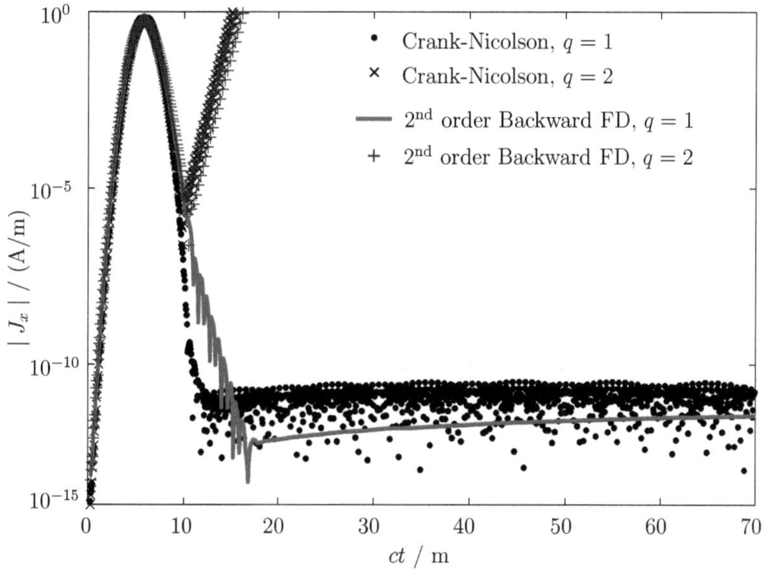

Figure 6.6: Magnitude of the induced surface current at the center of top face of the conducting cube, obtained solving the EFIE.

in the previous subsection. The effectivity of the proposed spline time bases is verified, as well.

The conducting cube is considered 1.0 meter on a side, divided monotonically into 4 divisions along the $\hat{\mathbf{x}}, \hat{\mathbf{y}}$, and $\hat{\mathbf{z}}$ directions resulting in 192 triangular patches with $M = 288$ common edges. The structure is illuminated by the $\hat{\mathbf{x}}$-polarized Gaussian plane wave propagating along the $\hat{\mathbf{z}}$ direction with the full-width half maximum of $\sigma = 0.7071$ lm (6.1). The governing DEFIE (2.11) is solved for the unknown induced surface current using the MOT procedure described in Section 3.3. It is observed that one can not proceed to such a small time step size, strictly speaking $\alpha \leq 2.7$, by the compatible integrator-interpolator pairs explained in Section 3.2, and yet expecting stable results over 30 lm-wide time interval. This is probably due to the fact that for small time steps the dominance of accurate evaluation of the time derivatives surpass over the effectiveness of the interpolation precision. The $\hat{\mathbf{x}}$-directed induced surface current at the center of the top face of the cube is shown in Fig. 6.7 where the second order accurate Crank-Nicolson integrator launches fluctuations about 12 lm after the start of simulations. Note that for the Crank-Nicolson integrator, (2.10) is solved using the linear Lagrange interpolation.

Fig. 6.8 illustrates the solution obtained using the new spline bases alleviate late-time instabilities until around 20 lm. Fig. 6.8 also exhibits that the resultant induced surface currents are in agreement with those obtained by the corresponding Lagrange approximations. Probing the magnitude of the current in logarithmic scale for a variety of α values reveals that the smoother Lagrange or spline functions are used, the later the exponential growth take places as shown in Fig. 6.9. This statement, however, is not generally true for

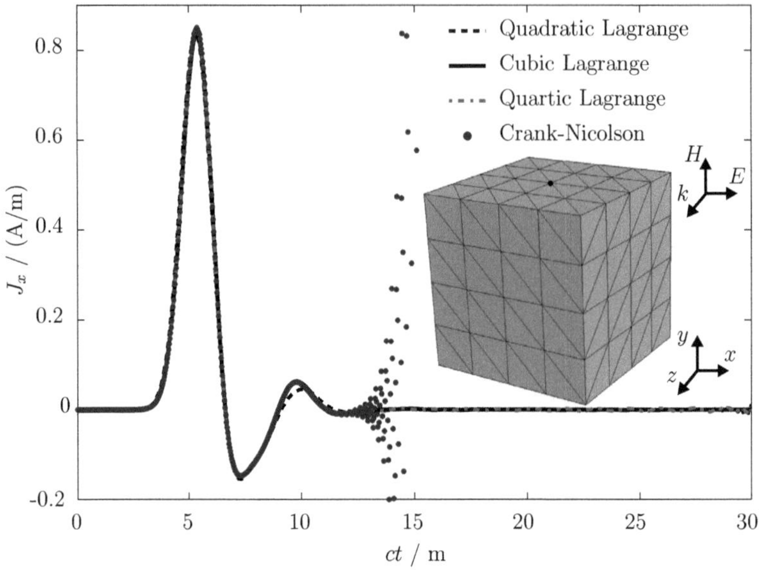

Figure 6.7: Transient induced surface current density at the center of the top face of a PEC cube 1 m on a side illuminated by the Gaussian plane wave using diverse time basis functions. The problem has been solved for the time step factor $\alpha = 0.99$ with respect to the smallest electric size of the spatial mesh.

other case studies.

As the second example, a flat 2 m×2 m square conducting plate of zero thickness located in the **xy**-plane and centered at the origin is considered. Ten and nine divisions are made along the $\hat{\mathbf{x}}$- and $\hat{\mathbf{y}}$-directions, respectively, resulting in 144 triangular patches with 251 common edges. In Fig. 6.10, the results obtained by the time shifted Lagrange basis functions and the most promising scheme proposed in Section 3.2 are compared. The corresponding plot in logarithmic scale is given in Fig. 6.11. It is observed that increasing the order of the interpolating Lagrange time basis functions does not necessarily postpone the occurrence of the exponentially growing late-time instabilities. Nonetheless, they all postpone the rise of the spurious oscillations later than those appear via applying the commonly used collocation techniques. Checking different combinations of the basis function orders (p,q) respectively for the time integration and interpolation, generally it is observed that using $p > q$ leads to stable results for smaller time steps than $p = q$ does. Although this coarse interpolation technique is in compliance with Section 3.2.3, it is not used in presenting the results here due to its filtering nature. Fig. 6.12 implies the results obtained using spline bases agree well with those of Lagrange bases. The absolute value of the $\hat{\mathbf{x}}$-directed induced surface currents at the center of the plate has been also shown in Fig. 6.13 in logarithmic scale so as to better demonstrate that the new spline time basis functions decay the amplitude of the late-time response on the desired solution interval pretty similar to the Lagrangian counterparts.

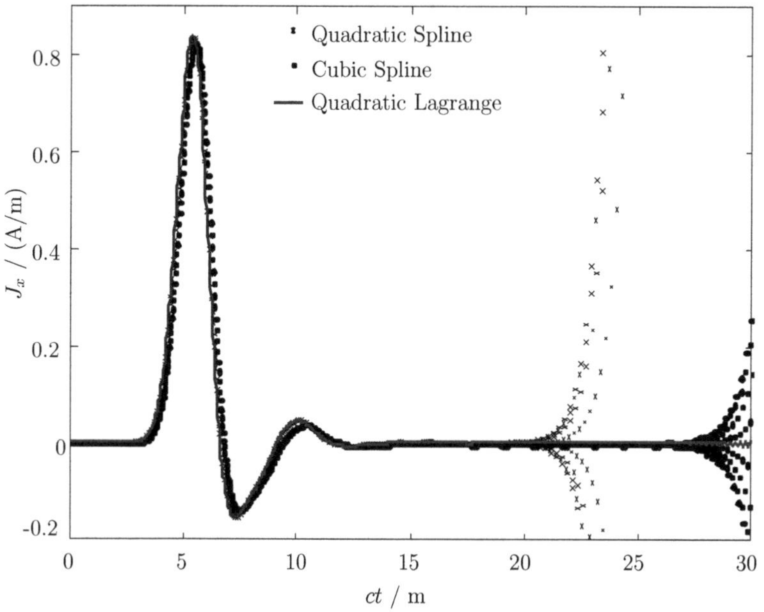

Figure 6.8: $\hat{\mathbf{x}}$-directed transient induced surface current density at the center of the top face of the 1 m conducting cube illuminated by the Gaussian plane wave, $\alpha = 0.99$.

6.4 Orthogonal Time Basis Functions

In this section, the recursive relations between different orders of the weighted Laguerre polynomials in (3.69)-(3.71) are considered for numerical solution of the TDIE. The Laguerre series is defined only over the interval from zero to infinity and, hence, are considered to be more suited for the transient problem, as they naturally enforce causality. In the MOD recipes, the only place where the time variable appears is the computation of the Laguerre polynomials of different orders approximating the incident pulse in (3.61) or (3.62) and/or (3.63). The accuracy of all explained routines in Section 3.4.1 was checked and for the sake of brevity here we only present the results of solving (2.11). Starting from (3.53) with the scaling factor $s = 1.0 \times 10^9$, for three order of Laguerre expansion $N = 20, 50$, and 80, system of equation (3.59), involving the separate spatial and temporal testing based on the Galerkin's method, is constructed to calculate the coefficients of the matrix equation (3.67).

Fig. 6.14 exhibits the convergence of the results for $M = 1152$ indicating that $N = 20$ is not sufficient to approximate the temporal variation of the response by the sum of Laguerre polynomials. Here, the temporal basis functions are completely convergent to zero as time increases to infinity. Therefore, the transient response spanned by these basis functions is also convergent to zero as time progresses and thus there is no necessity to develop the BEM in the CFIE arrangement as [21]. Fig. 6.15 reveals that the higher the last order of the MOD method N is, the better the late-time solution decays.

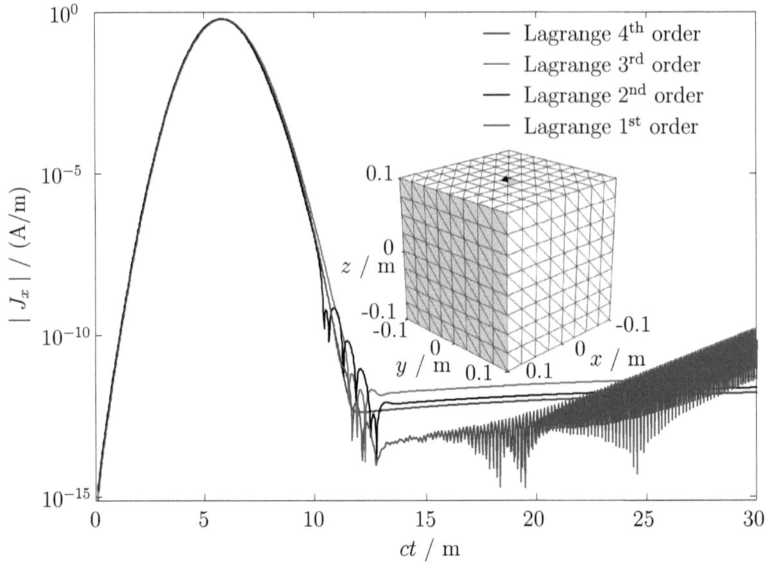

Figure 6.9: $\hat{\mathbf{x}}$-directed transient induced surface current density at the center of the top face of the 0.2 m conducting cube obtained in $\alpha = 1.7$ by different orders of the interpolating Lagrange time basis functions.

6.5 Hybrid Meshes

The second-order Lagrange (quadratic) interpolants (3.37) are chosen as the temporal BF. The time step size is chosen according to a fraction of minimum distance between the centroids of subdomains in light meter (lm), i.e., $\Delta t = \alpha \frac{R_{\min}}{c}$. First, a conducting pie-shaped plate is illuminated by an x-polarized Gaussian plane wave propagating along z-direction with the full-width half maximum of 1.4142 lm, according to Fig. 6.16. Based on the generatrix definition in Section 2.4.4, the generatrix here is a straight line parallel to the y direction, generating nodes along the x axis. To match the transversal coordinate of the boundary nodes on the diameter of the semicircle with those of the connected parts, a pie-shaped premodel consisting of the tapered sector and non-coplanar semicircle (rotated θ degree around the diameter) is first imported to the automatic triangular mesh generator. After triangular mesh generation, the semicircle nodes of the output are then rotated back to the xy-plane. Finally, to make the composite mesh, the strip transitions are placed between the leading and trailing edges of the rectangular block. The initial numbering of the edges shared between the border triangles and newly inserted rectangles are modified only for the leading rectangle edges and updated for the trailing triangle edges connected to the end of the generatrix so as to respectively match and adopt the edge numbering for the unified structure. The TD EFIE is solved by the implicit MOT scheme with the time step factor $\alpha = 2.1$. The x-directed induced current density at the center of the generatrix is shown in Fig. 6.16. The results obtained using the triangular mesh exclusively ($M = 453$ RWG BF) are in good agreement with the composite mesh ($M = 360$ hybrid BF).

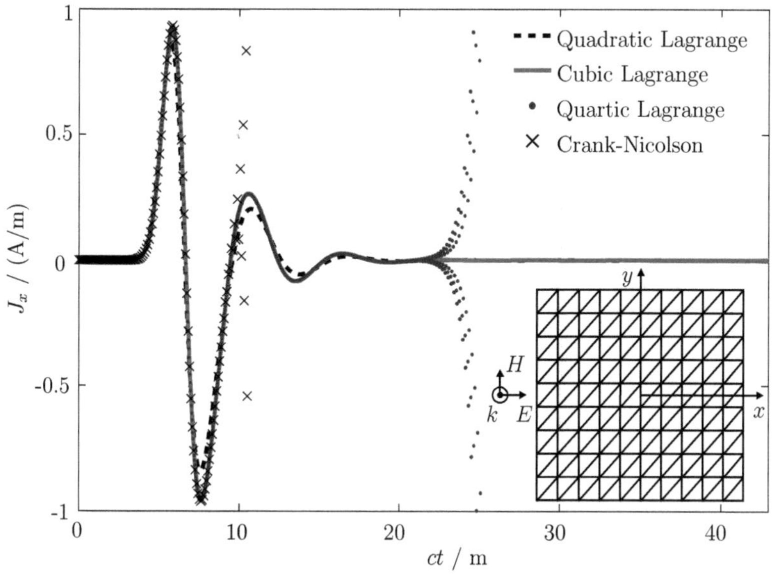

Figure 6.10: Transient induced surface current density at the center of the 2 m ×2 m conducting plate illuminated by the Gaussian plane wave with $c\sigma = 0.7071$ m using diverse time basis functions. The time step factor $\alpha = 0.9$ has been chosen.

Fig. 6.17 exhibits a hollow tube excited by an incident Gaussian plane wave with the full-width half maximum of 0.3535 lm and discretized half by the RT BF and half by RWG BF. Fig. 6.18 depicts diverse models of a pillbox cavity [55] consisting of three circular generatrices. The generatrices can be shifted for creation of cylindrical parts meshed by rectangles. In other words, the edge vertices lying on the circular interfaces of cylindrical parts are longitudinally extended and rotationally replaced by the parallelogram grid nodes to constitute the polyhedrons in between, i.e., the positions of the inserted nodes lie on the radial and the azimuthal coordinates ρ and ϕ = const. of the interface nodes and $z_L < z < z_R$ between the left and right cutting planes. A bunch of electrons with the total charge of $q = 1$ nC passes through the cavity along the z-axis with the speed of light c. The particles are modeled by a line charge[1] with normalized Gaussian spatial distribution and the full-width half max $\sigma = 0.025$ m equals one-forth of the largest dimension of the body. Referring to Fig. 6.18, (a) and (e) infer the unitary mesh completely generated by the MWS software package, (f) and (d) indicate the RT BF on double-side tune arms, and (b) and (c) point to the fully hybrid planar mesh cases, respectively, for the closed and open models. The incident field is zero when the charged particles are outside of the open or closed cavity. The time step size is set by $\alpha = 2.5$ for pure triangular meshes and $\alpha = 4$ for hybrid cases. The total surface (the induced plus the image) current density obtained using the hybrid BF and RWG BF alone are compared for the closed models in Fig. 6.19. The z-directed total current density at the middle of the open structures is

[1]For further explanations, the readers are referred to Section 6.11.

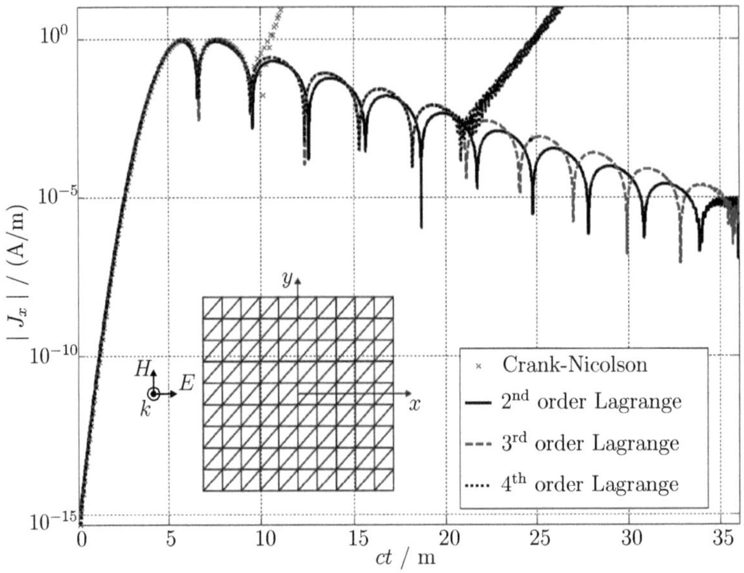

Figure 6.11: Transient induced surface current density in logarithmic scale at the center of the conducting plate illuminated by the Gaussian plane wave using diverse time basis functions, $\alpha = 0.9$.

shown in Fig. 6.20. The difference in the results after the bunch leaves the structure is due to the generation of slightly different higher modes by the different coarse meshes. Since the radial incident field only affects cylindrical walls and the excitation starts when the bunch arrives to the structure, the response of the open and closed models are very similar as the comparison of Fig. 6.19 and Fig. 6.20 reveals.

Perpendicular edges of rectangular meshes on generatrices directly gives the two orthogonal ($\hat{\mathbf{z}}$,$\hat{\phi}$) components of the current. Negligible values of the azimuthal component of the current (longitudinal component of the tangential field) additionally asserts the validity of simulations. Smaller amount of the azimuthal component of the current is observed in the hybrid cases. The meshed domain in the BEM is confined to the surface of structures, and hence, it does not provide a realistic unbounded movement for the induced image charges by the wake fields travelling backward. As a result, the image charges are reflected at the ends of the open or closed arms, and thus, the simulation is valid until the image charges front reaches to the behind tube end, somewhere between the first zero padding and the first undershot.

The range of α ensuring stability of the algorithm depends on the geometrical discretization of the body [5]. The more uniform and regular is the combined patch modeling of the scatterer, the wider is likely to be the set of values of time steps leading to a stable behavior for the MOT by the hybrid bases. The uniformly meshed cylindrical parts by the rectangles allows one to directly take advantages of the translational invariance property of the Green's function in the MOT's convolution products, Section 6.7. Fig. 6.21, Fig. 6.22,

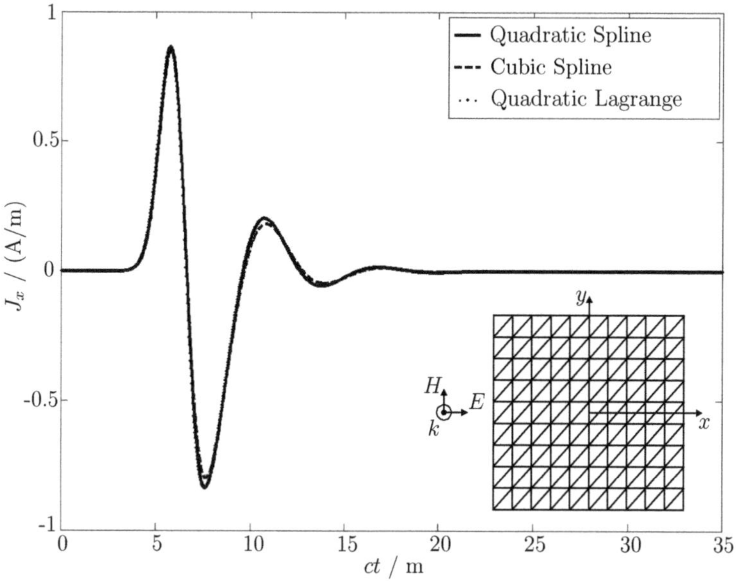

Figure 6.12: $\hat{\mathbf{x}}$-directed transient induced surface current density at the center of the $2\,\mathrm{m} \times 2\,\mathrm{m}$ flat plate, $\alpha = 0.9$.

and Fig. 6.23 exhibit diverse application of hybrid meshes generated through replacement of triangles by rectangles on generatrices.

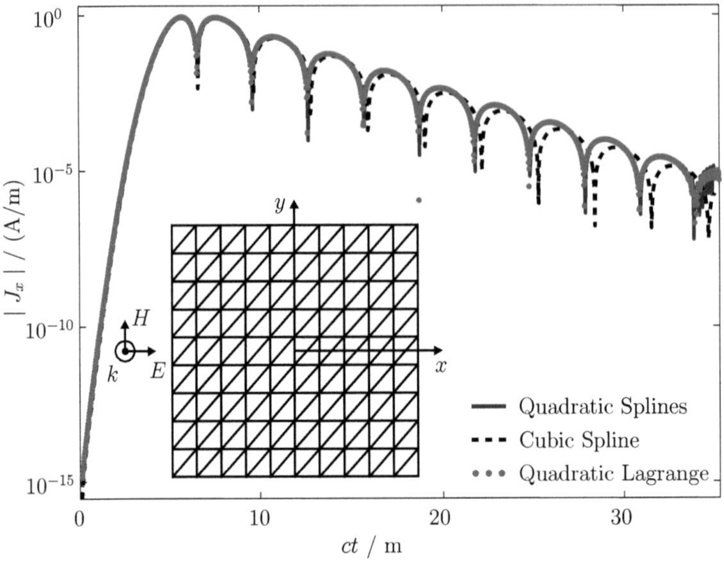

Figure 6.13: Magnitude of the current density at the center of the PEC plate in logarithmic scale, $\alpha = 0.9$.

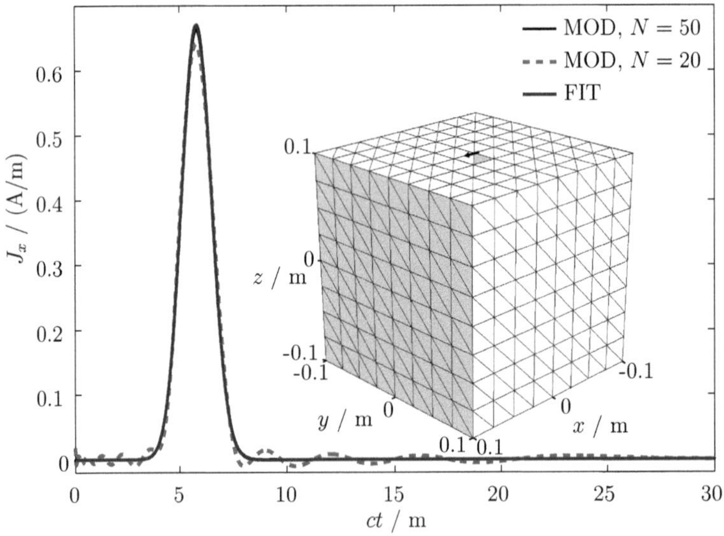

Figure 6.14: $\hat{\mathbf{x}}$-directed transient induced surface current density at the center of top face of the conducting cube using the MOD method.

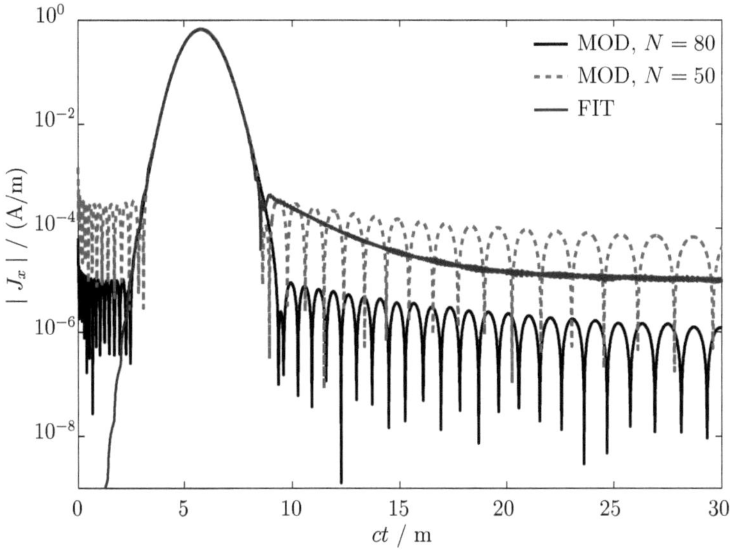

Figure 6.15: Magnitude of the current density at the center of the PEC cube, Fig. 6.14, in logarithmic scale.

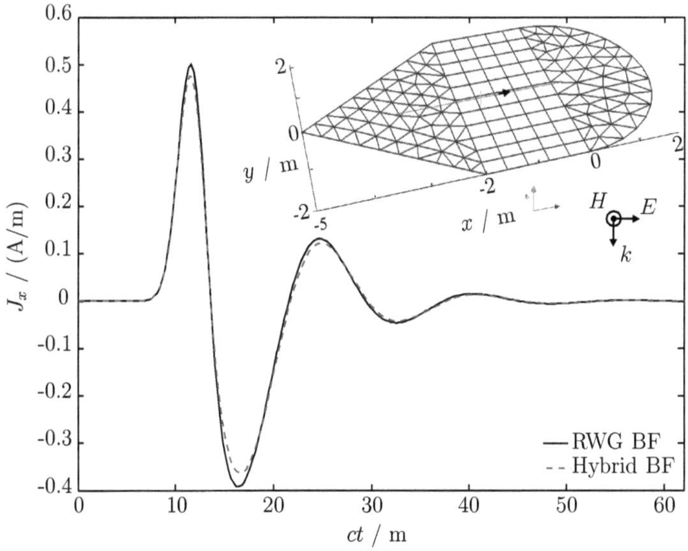

Figure 6.16: Current at (-1,0,0) the stretched pie-shaped discretized by RWG-RT BF.

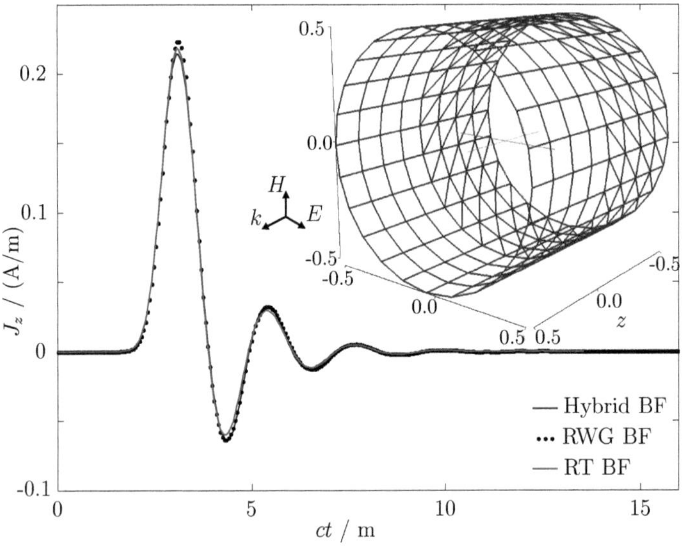

Figure 6.17: Current on the mid top of the tube discretized by RWG-RT BF.

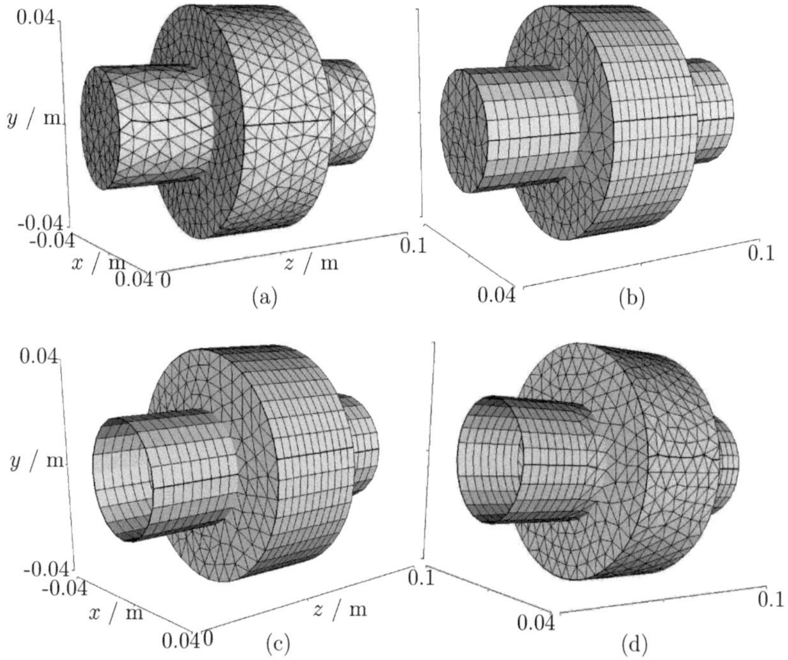

Figure 6.18: Surface discretization choices for a pillbox (cylindrical) cavity model including beam tubes, and the associated number of unknowns: (a) RWG BF over the entire body, $M = 2940$. (b) RT BF on all three cylindrical parts, $M = 2470$. (c) Uniform extension of beam pipes in longitudinal direction, $M = 2361$. (d) RT BF only on tube arms far from the sharp wedged corners, $M = 2656$. Additionally, the surface mesh is generated for the cases: (e) Open duplicate of (a), $M = 2518$ and (f) Closed counterpart of (d), $M = 3076$. The rectangular patches on the cylindrical beam tubes are generated by transferring the interface circles of the corresponding premodel in longitudinal direction.

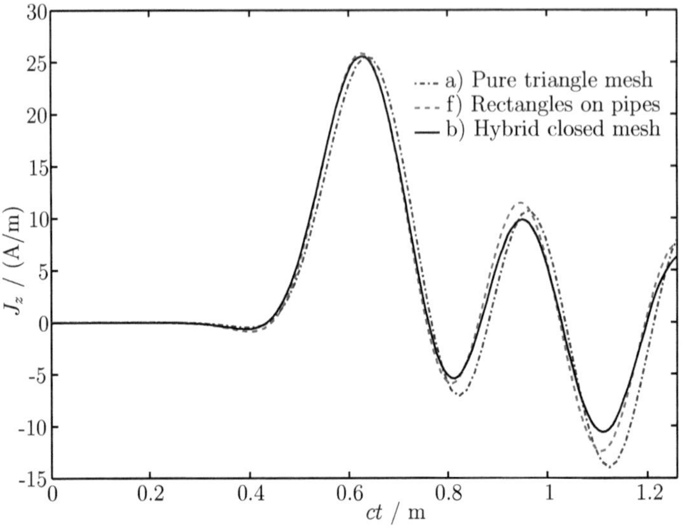

Figure 6.19: Total current on mid top of the closed pillbox cavity models in Fig. 6.18. (a) unitary triangular mesh, (b) rectangular mesh on cylindrical parts, and (f) rectangular mesh only on beam pipes.

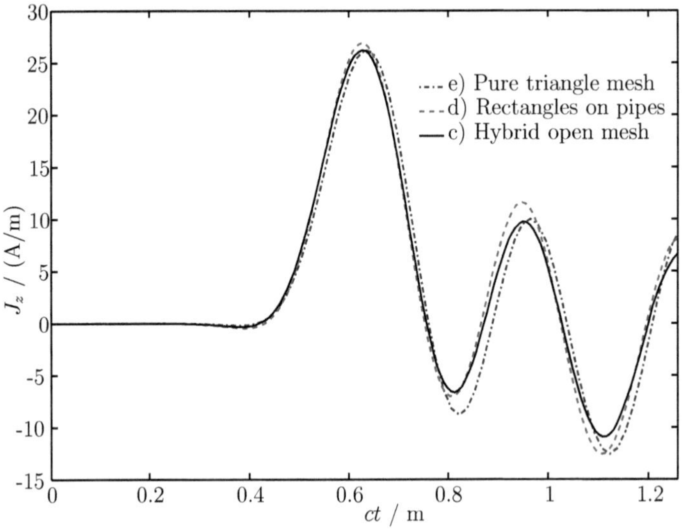

Figure 6.20: Total current on mid top of the open pillbox cavity models in Fig. 6.18. (e) unitary triangular mesh, (c) rectangular mesh on cylindrical parts, and (d) rectangular mesh only on beam pipes.

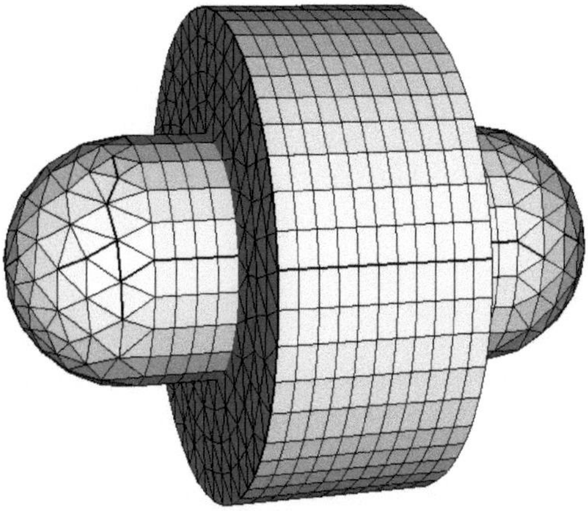

Figure 6.21: A hybrid mesh model for an elliptical pillbox cavity [125] with beam pipes enclosed by hemispheres.

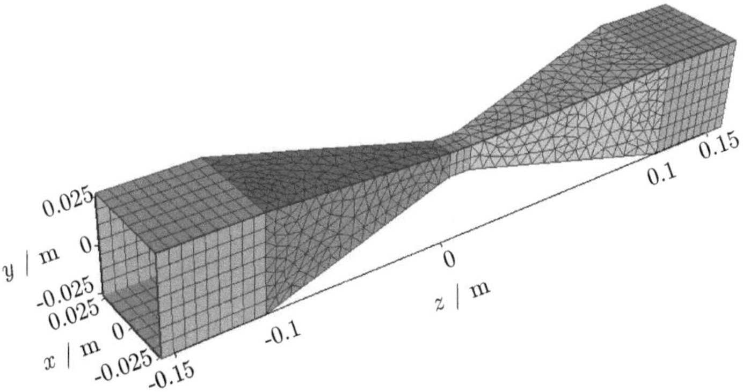

Figure 6.22: Rectangular collimator. Generatrices are automatically meshed by rectangular patches.

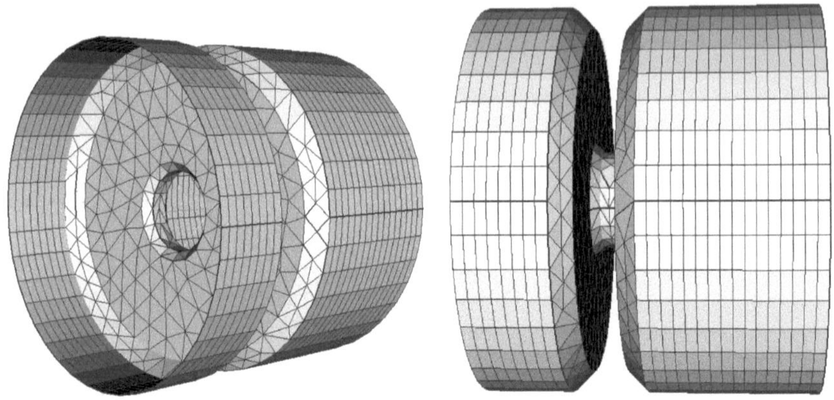

(a) (b)

Figure 6.23: An open model for the RF Gun in PITZ cavity. The triangular mesh is generated by an automatic surface mesh generator. The tube arms are extended by adaptive plantation of rectangular patches.

6.6 FDDM and CQM

The FDDM and CQM methods described in Section 3.5 are first applied to the plane-wave scattering from the PEC cube. The conducting cube is considered 0.2 m on a side, divided monotonically into 4 divisions along the $\hat{\mathbf{x}}, \hat{\mathbf{y}}$, and $\hat{\mathbf{z}}$ directions resulting in 192 triangular patches with $M = 288$ common edges. The representative structure is illuminated by the $\hat{\mathbf{x}}$-polarized Gaussian plane wave propagating along the $\hat{\mathbf{z}}$ direction with the full-width half maximum of $\sigma = 0.7071$ lm (6.1). The governing DEFIE (2.11) is discretized in the spectral domain for the unknown induced surface current using the CQM procedure described in Section 3.5.1. It is observed that one can not proceed to very small time step size, strictly speaking $\alpha \leq 0.6$, by the first-order spatial basis functions and yet expecting stable results. The $\hat{\mathbf{x}}$-directed induced current density at the center of the top face of the cube calculated in different sampling-rates is shown in Fig. 6.24. The convergence of the $\hat{\mathbf{x}}$-directed induced surface current to the FIT reference solution is observed for small temporal sampling rates $\Delta t = \alpha \frac{R_{\min}}{c}$. Comparison of the results for the fine mesh in Fig. 6.24 with those of the coarser one (e.g., in Fig. 6.44(a)) reveals that the existing time shift due to different sampling rates vanishes gradually by refining the space mesh so that the FDDM results converge to the FIT reference solution. To better visualize the robust stability of the FDDM for the broad sampling rate choices, the absolute values of the responses are plotted in logarithmic scale in Fig. 6.25.

Fig. 6.26 expresses good agreement between the higher-order FDDM and the MOD results for the PEC ogive illuminated by $\sigma = 3.5355$-width incident pulse, even when only the early computed $N_g = 50$ interaction matrices are employed, i.e. $\bar{\bar{\mathbf{Z}}}_i \simeq 0$ is assumed for $i > 50$. Analysis of a nonuniformly-meshed trihedral corner reflector in Fig. 6.27 indicates that the properly chosen $N_g \geq 33$ for eliminating the almost-zero late-appearing matrices from the construction of the right side of the matrix equation do not harm at all the stability and accuracy of the transient analysis. Note that the transient currents were obtained by adaptive refining quadruple quadratures. For the numerical evaluation of the vector and scalar potentials, respectively, four and seven-point quadrature methods were used on (partitioned) spatial subdomains.

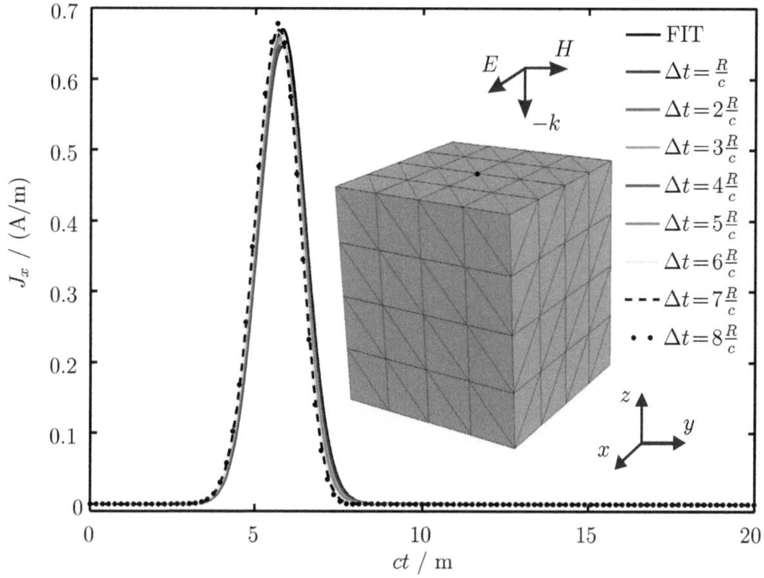

Figure 6.24: $\hat{\mathbf{x}}$-directed surface current density at the center of the top side of the PEC cube, $M = 288$.

6.7 Space-FFT Acceleration on Uniform Meshes

An electrically large dipole antenna in receiving mode with a total length of $L = 40$ m and circular cross section with a radius of $a = 1.0 \times 10^{-3}L$ is considered. The wire is modeled by a narrow strip [10, 43] and approximated by $N_s = 100$ RT basis functions. As Fig. 6.28 implies, the space-FFT approach in Section 4.2 reduces the computational complexity and memory requirement for plenty of (4.4) without diminishing the accuracy or disturbing the stability. Simulation results of the 100-wavelength long strip in Fig. 6.28 also affirm that the algorithm (4.1) is aliasing free. Note that misalignment of even one entry during the FFT immediately causes explosion of the response amplitude.

A second example is given by a flat conducting plate of zero thickness in form of a 1 m×1 m square. It is located in the xy-plain and centered at the origin. With the choice $N_x = 5$ and $N_y = 4$, respectively, six and five divisions are made along the x and y directions which results in total $N_s = 49$ common edges. The governing TD EFIE (2.10)

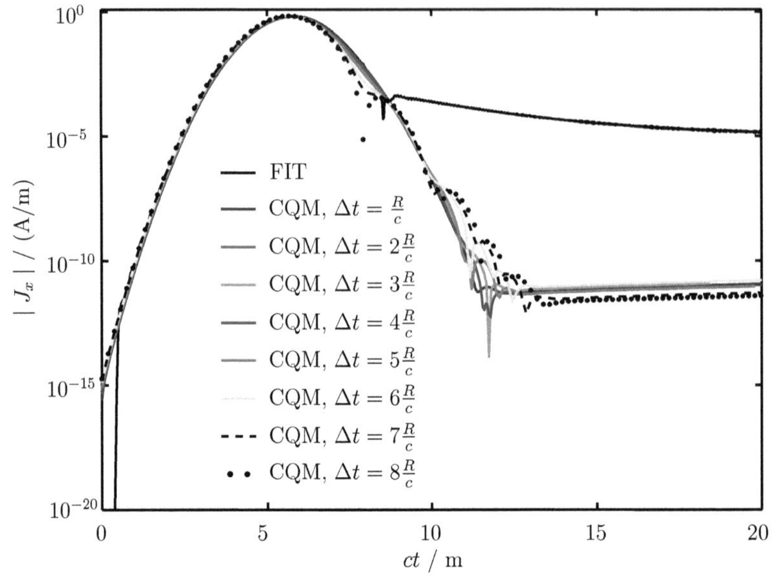

Figure 6.25: Magnitude of the current density at the center of the top side of the PEC cube (Fig. 6.24) in logarithmic scale.

is solved by the AMOD with $s = 3 \times 10^8$ for $N_t = 80$. Fig. 6.29 shows the induced current in the middle of the plate. Continuing the marching process for the higher order Laguerre polynomials ($N_t > 120$), the unphysical early ripples are totally vanished. As Fig. 6.29 illustrates there is a good agreement between the results of the introduced AMOD scheme and its FFT-accelerated version in which every full matrix-vector multiply is replaced by a convolution with a reversed vector, which is implemented through a double precision outer product in the Fourier domain (4.1).

Considering the plate meshed with $N_x = 11, 13, \cdots, 21$, and $N_y = N_x - 1$ divisions, the computing times of the marching process for the AMOD method up to $N_t = 80$ and the MOT algorithm till $N_t = 100$ are plotted in Fig. 6.31 versus the number of spatial unknowns. As can be deduced from the figure, the fully populated matrices of the MOD method benefit from the space-FFT approach more markedly than the sparse matrices of the MOT method do.

Fig. 6.32 investigates the impact of wavelet packet basis transform on all interaction matrices to further take advantages of sparse matrix storage in the MOT. The decomposition tree is designed based on minimizing the cost function for the coefficient matrix $\bar{\bar{Z}}_0$ as explained in Section 4.5.1. After the fast wavelet transforms, those matrix elements smaller than $\epsilon = 0.1 \times 10^{-3}$ factor of the matrix $\bar{\bar{Z}}_0$ energy (second norm) are omitted. The average sparsity in wavelet domain reaches to 77 percent which is few percents higher than that can be reached by direct thresholding in space domain.

As the next example, a 0.5 m long hollow tube with radius 0.1 m is uniformly meshed by ($N_\phi = 18$, $N_z = 20$) rectangular planar patches and is analyzed by the MOT scheme

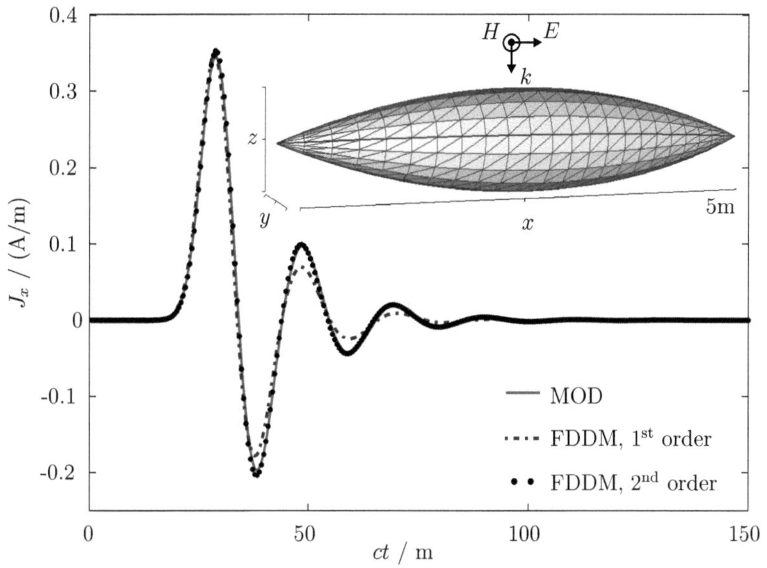

Figure 6.26: **x**-component of the induced surface current density at the middle of the ogive PEC with extruding angle 22.62°; The FDDM solution using the second order BFD converges to the MOD solution; $M = 1188$, $\alpha = 18$.

with $N_s = 666$ rooftop bases for $N_t = 150$. Fig. 6.33 demonstrates that using the FFT algorithm proposed in (4.6) for $N_g N_t = 900$ matrix-vector multiplications does not degrade the accuracy or perturb the late-time stability of the response.

6.7.1 Finite Periodic Structures

Fig. 6.34 shows a 5 × 5 finite-sized array of 1 m×1 m patches separated by 0.5 m. For the doubly finite periodic distribution of the antenna array, the algorithm exploits the block-Toeplitz structure of the interaction matrices in four nested levels to multiply all matrix-vector by the FFT, $n_x = n_y = 5$, $N_x = 7$, $N_y = 6$, $N_s = 2425$ in Fig. 4.3. The four nested Toeplitz levels (while the outer two levels are dealt with the periodicity) totally reduce the computational effort to $\mathcal{O}(N_t^2 N_s \log N_s)$.

6.8 Reduced Sum Convolution Products

The validity of the improved schemes in Section 3.4.4 in proper modeling of the induced surface current density was checked for several three dimensional PEC objects under the plane-wave incidence. For the sake of brevity, three case studies are presented here, including the next related Section 6.9.

A PEC sphere with radius 0.25 m centered about the origin is approximated by 472 triangular patches with $M = 708$ common edges. The object is illuminated by an x-

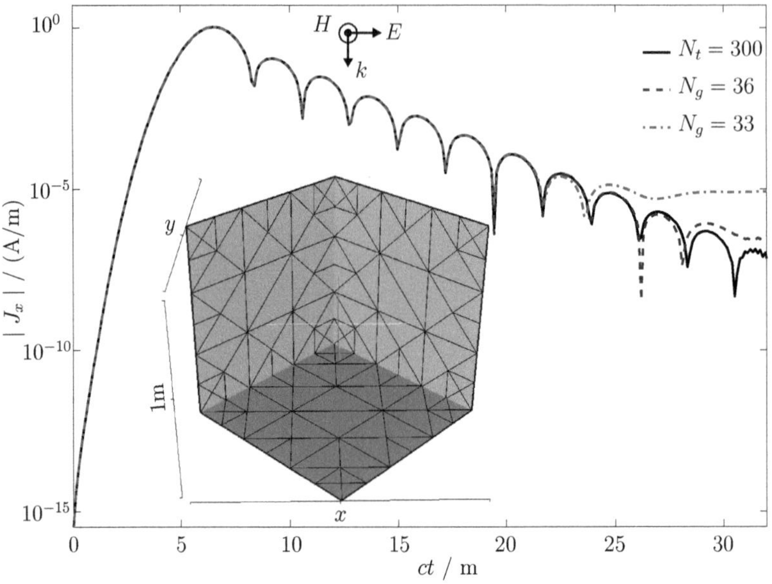

Figure 6.27: The effect of the proposed cut-off strategy.

polarized incident Gaussian-shaped plane wave with the full-width half maximum of $\sigma = 0.7071$ lm, Fig. 6.35. The governing TDIE (2.10), (2.11), and (2.24) are solved by the AMOD with $s = 1.0 \times 10^9$ for $N = 50$. Fig. 6.35 exhibits the electric current density J_z on the sphere surface at $\mathbf{r} = (0.25, 0, 0)$ obtained using the diverse MOD schemes equipped with the introduced summation reduction techniques. Figure 6.36 shows agreement with the results obtained by the MOT counterparts using the subdomain quadratic Lagrange interpolatory time bases with $\Delta t = 3.5\frac{R_{\min}}{c}$. The results also matches well to the high-resolution finite integration technique (FIT) [126]. Summation reduction when taking the time derivative from both sides of the MFIE (DMFIE) was also successfully tested. Except in the expansion by the combination of one past, present and one future Laguerre polynomials in (3.99), the other summation reduction techniques do not harm the high accuracy of the AMOD methods at all. Result agreements were also obtained when the conventional MOD approximation (3.65) is used.

6.9 Time-FFT Speed Up

In Table. 6.1 the computational time of the accelerated algorithms using Toeplitz property on time steps for $N = 2^{13}$, $M = 1$, and $B = 128$ is compared. Although the experimental speed-up factor when the varying-size blocks smaller than $B = 128$ are excluded from the FFT approach (signified by Fig. 4.4△) is quite notable in Table. 6.1, once the algorithms are combined by the generic spatial convolutional products, the middle columns' values grow to the extent that it worths to use the FFT also for the small early matrix blocks and

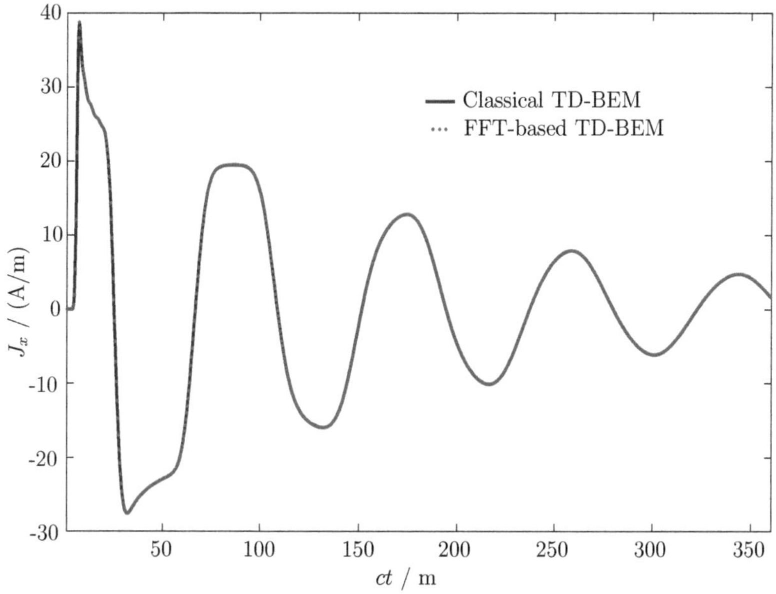

Figure 6.28: Transient electric current density at the mid point of a 40 m-long narrow strip under the plane-wave incidence.

Table 6.1: CPU run time in second for different convolutional strategies in performing the block aggregate matrix-vector multiplies in marching schemes.

Conventional product	FFT-based Multiply Algorithms			
	Fig. 4.4	Fig. 4.4△	Fig. 4.5	Fig. 4.6
6.45	2.25	0.4	0.39	2.13

the computational efficiency of the last grouping method in Fig. 4.6 outrun the others.

The block aggregates of interaction matrices and the sequence of current vectors are shifted one row block up-left and one upward, respectively, to entirely adopt the present FFT algorithms for the (A)MOD methods specifically. The next example is a cone sphere with the total length of 1.235 m approximated by 1044 triangular patches with $M = 1566$ as shown in Fig. 6.37. The governing DEFIE (2.11) is solved by the time FFT-accelerated AMOD with $s = 2.0 \times 10^9$ for $N = 128$ and $B = 32$. The matrices $\bar{\bar{Z}}_\nu$ are symmetric when the double surface integrals are calculated in a totally symmetric way [127], that is the number of quadrature points are adaptively refined by simultaneous partitioning of the source and observation subdomains. The FFT is applied to the blocks only in the first calling, as illustrated by the solid boxes in Fig. 4.4 and Fig. 4.5. Since in the FFT approach larger block aggregates are multiplied less frequently, the total computational complexity of the MOD methods is reduced, especially for large time-bandwidth product N. The complexity reduction in temporal dimension does not depend on the spatial mesh.

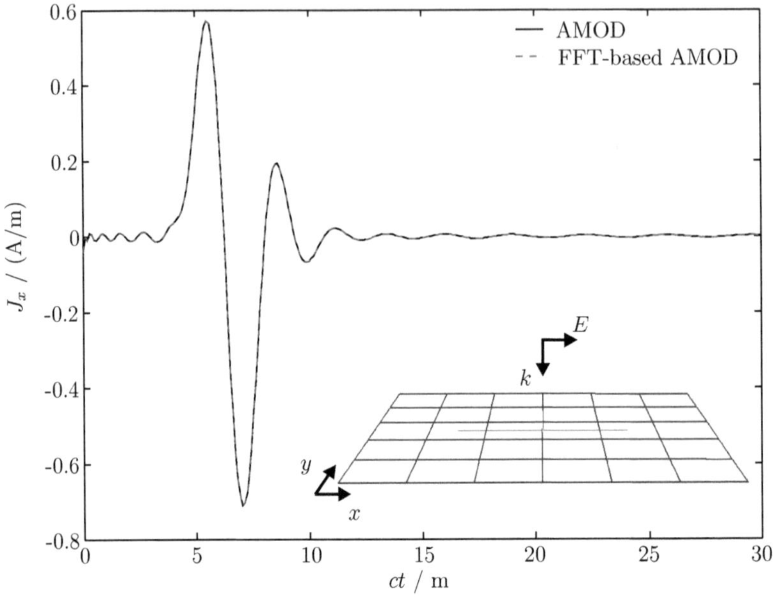

Figure 6.29: Induced surface current density at the center of the conducting sheet.

As the last example, a PEC NASA almond with length 1.0 m approximated by 864 triangular patches with 1296 common edges is illuminated by an x-polarized incident Gaussian-shaped plane wave with the full-width half maximum of $\sigma = 1.0$ lm, Fig. 6.38. The governing TD EFIE (2.10) is solved by the time FFT-accelerated MOD with $s = 2.0 \times 10^9$ for $N_t = 128$ and $B = 32$. The FFT is applied to the blocks only in the first calling. Fig. 6.38 exhibits the electric current on the middle of the top of the almond surface obtained using the MOD scheme equipped with the introduced summation reduction technique. Figure 6.38 also shows agreement with the results obtained by the MOT counterpart using the subdomain quadratic Lagrange interpolatory time bases with $\Delta t = 9\frac{R}{c}$. In the FFT-based MOT, the largest temporal FFT size is proportional not to the duration of the analysis but to the maximum transit time across the scatterer, N_g.

6.10 Polynomial Eigenvalues of TDIE Solvers

The assembly of the (artificial) iteration matrix for the TDIE solvers has been explained in Section 3.7. Fig. 6.39-6.45 demonstrates the scattering response of different unit objects under a Gaussian plane wave incidence with the full-width half maximum of $\sigma = 0.7071$ lm as well as the eigenvalue locus of the iteration matrix for diverse marching schemes. Fig. 6.39(a) explicates some large negative eigenvalues for the system (3.140) residing outside the unit circle in the complex plane which cause growing oscillations that alternate in sign at each time step. The system eigenvalues lying outside the unit circle do not cause instability for the MOT as long as the corresponding frequency components are not excited

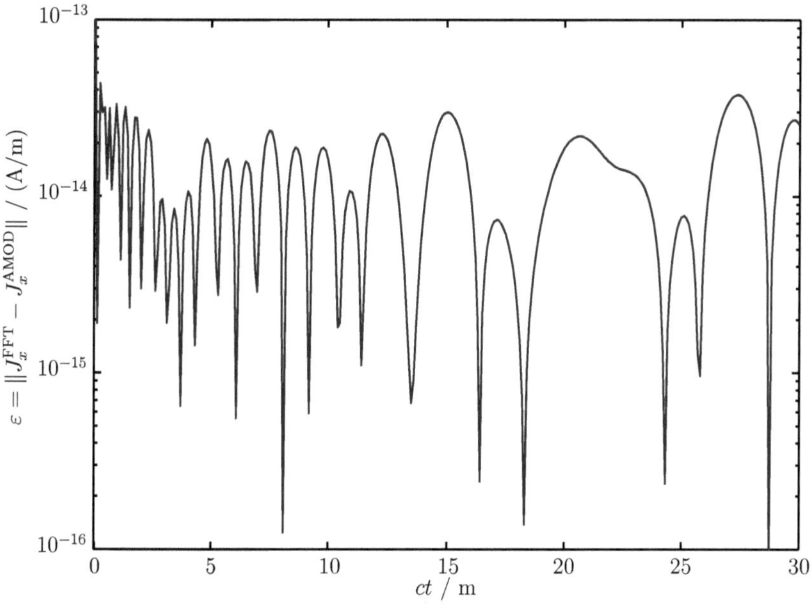

Figure 6.30: Relative error due to the FFT-based matrix-vector multiplications in the AMOD result depicted in Fig. (6.29).

by the incident field. Using larger time steps in the MOT schemes although dwindles the hazardous eigenvalues located far left out of the unit circle, sinks simultaneously the inner ones to the center. Similar to the often applied filtering of high frequency modes by an averaging technique [1, 7, 13, 14, 16, 18, 19, 23], this contraction although expands the stable region, in fact it postpones the late-time instabilities by "pulling more energy out" of the system. Fig. 6.39(b) similarly investigates the scattering analysis of the unit sphere by adaptive partitioning of the subdomains to retain the relative quadrature error less than $\epsilon_r = 0.1 \times 10^{-3}$. The simultaneous subdivision of source and test subdomains for symmetric calculation of the double surface integrals does not really have a considerable impact neither on preserving the energy content of the system nor on the condition number of $\bar{\bar{\mathbf{Z}}}_0$ or $\bar{\bar{\mathbf{Z}}}$. The same is observed with synthetic symmetrization of the matrices $\bar{\bar{\mathbf{Z}}}_n^{\mathrm{sym}} = \frac{1}{2}(\bar{\bar{\mathbf{Z}}}_n + \bar{\bar{\mathbf{Z}}}_n^{\mathrm{T}})$, where T denotes the transpose of the matrix. The nonsymplectic MOT methods dissipate the energy exponentially out of the system. The observation of late-time stability in the results, however, is not only due to the intrinsic damping of the time integrator. Using very small time step excites the modes outside the unit circle as the spectrum content of the incident Gaussian field covers the associated frequencies.

Fig. 6.40 shows MOT iteration matrix eigenvalues in solving the MFIE. Fig. 6.40(a) considers a cube scatterer to demonstrate that when the smaller the time step $\Delta t = \alpha \frac{R_{\min}}{c}$ is chosen, as the more simulation steps N_g takes for the EM fields radiating from source patches to leave the structure, the MOT algorithm has less dissipation. Fig. 6.40(b) considers a sphere scatterer to illustrate that the finer the mesh (the more number of space unknowns M) is used, the MOT dissipates less energy out of the system at the same time

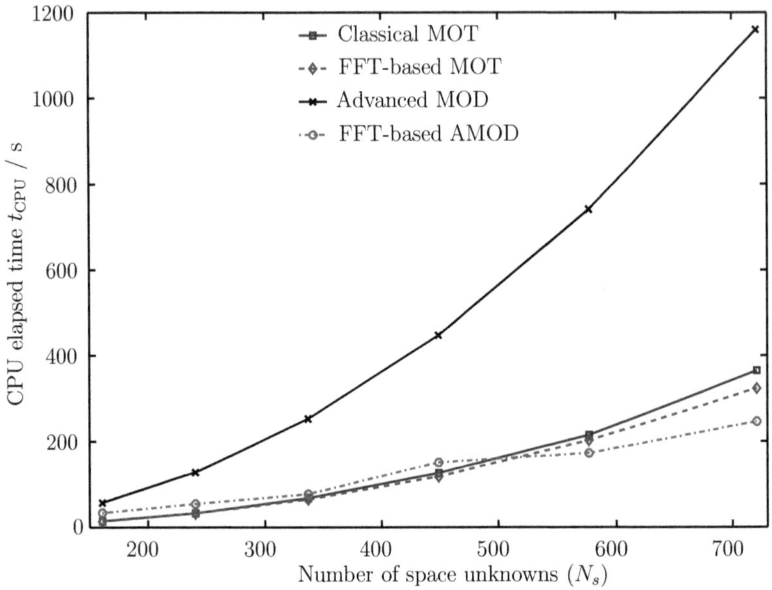

Figure 6.31: Average marching time in the AMOD and MOT methods versus the FFT-based counterparts for Fig. 6.29. The horizontal axis is the number of spatial unknowns.

step size. Fig. 6.41 exhibits the stabilization effect of the linearly weighted averaging in the CFIE on the dissipative MOT scheme. The cube and sphere scatterers in Fig. 6.41(a) and Fig. 6.41(b) are discretized respectively with 288 and 708 RWG bases where 4 and 6 past impedance matrices contribute in the solution procedure. Thus, when EFIE is solved by the classical MOT, system eigenvalue(s) locate out of the stability unit circle, for the MFIE and CFIE, however, this is not the case.

Fig. 6.42(a), 6.42(b), with respectively $N_t = 40, 100$, exhibit the eigenvalue distribution of the commonly used MOD recipes in which the outer surface integrals over the observer patches are approximated by the value of the respective integrands at the centroid of the triangles (3.65). No degenerated DC mode exists near $\lambda = 1 + j0$ in the plots. As Fig. 6.42(b) elucidates, in contrary to the MOT in the late time, the larger the scaling parameter (the finer the time resolution) is in the MOD, the less susceptible the system is to immensely large eigenvalues leading to ripples at early times. Although the eigenvalues are not totaly reside on/in the unit circle, no spurious fluctuation is observed at the late-time since the energy content decays exponentially by the damping weighted coefficients of the Laguerre expansion. Thus, the Galerkin testing on orthogonal time bases is necessary but not a sufficient condition for symplectic integration. The essence of extremely large eigenvalues is attributed to the unrealistic assumption of no changes for the unknown transient quantity within the subdomains (3.8). Similar results are obtained for the MFIE. Fig. 6.43(a) considers the radiation of a dihedral corner reflector and 6.43(b) a hollow hemisphere using the advanced MOD (AMOD) methods. Fig. 6.43 reveals that the AMOD methods are highly immune to the numerical inaccuracy due to the energy dissipation,

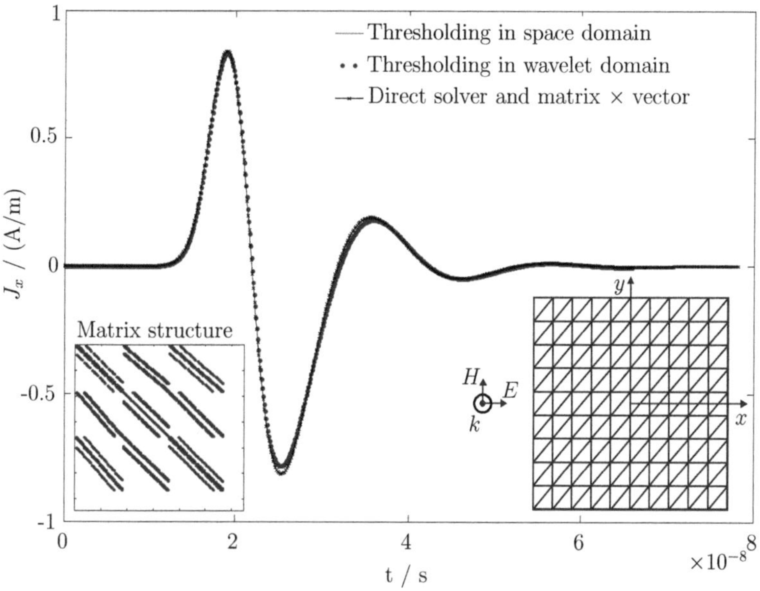

Figure 6.32: $\hat{\mathbf{x}}$-directed induced surface current density at the center of the 2 m ×2 m PEC plate illuminated by the Gaussian plane wave with $c\sigma = 0.7071\,\mathrm{m}$ using sparse storage after wavelet packet basis transform on all interaction matrices.

provided that sufficiently large number of Laguerre expansion orders are proceeded. By the virtue of using the AMOD formulations, the eigenvalues of the delay integrodifferential equation system merge to the unit circle regardless of the time resolution (scale parameter s) value. Likewise, the Hermite polynomials can also be adopted to solve the TDIE so as to generate an alternative nondissipative scheme, Section 3.4.6. Once the outer space integrals in the AMOD are approximated by the value of the integrands at single points on the barycenter of the observation patch, similar behaviour in the pole displacement is observed in Fig. 6.43. Hence, performing symmetric quadratures for the spatial discretization is not necessary for energy conservation. In fact, to render an energy conserving space-time integration for the TDIE solvers the reciprocal time testing and the inclusion of the retarded distance in the source integrals, as the AMOD does, are necessary. The reciprocal space Galerkin method is also necessary [86] but the precision of evaluating the spatial testing integrals is not so important and the value of the outer integrand can be approximated at the barycenter of the observation patches.

Fig. 6.44 and Fig. 6.45 exemplify that the eigenvalue distribution of the FDDM system transfer function consists of a circular cloudy quasi-ring. The FDDM system matrix always has a DC mode at $\lambda = 1 + j0$ and no exact zero eigenvalue $\lambda = 0$ (are not singular). Fig. 6.44(a) considers solving the scattering problem of a PEC cube with a side length of 0.2 m meshed by $M = 12$ triangles at different time steps for $N_t = 200$. It displays that the eigenvalue locus of the FDDM system heads toward the unit circle in the complex plane for small sampling rates. As seen in Fig. 6.44(b), also by marching on more samples, poles

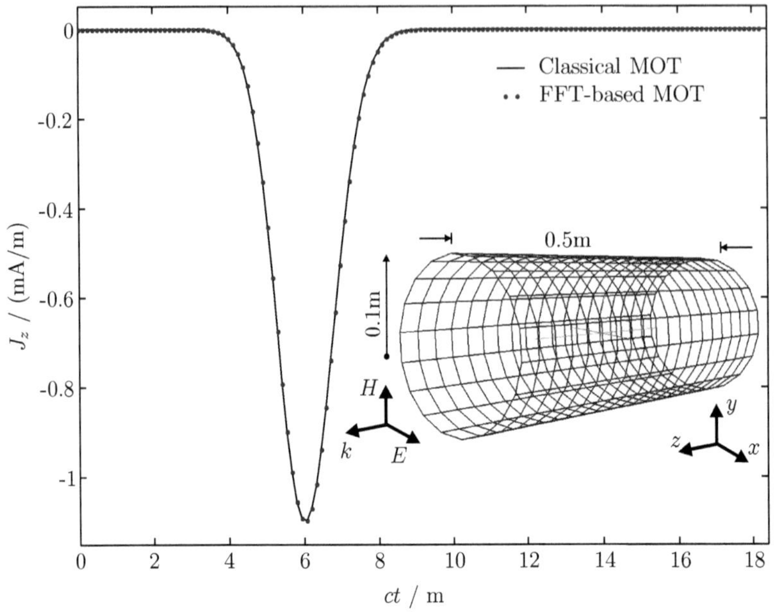

Figure 6.33: Transient electric current density on the middle of the tube surface under plane wave excitation, $\vec{\mathrm{k}}||\hat{\mathbf{z}}$.

take distance from the origin so as to ultimately approach to the unit circle for sufficiently large number of time-frequency sampling points. Fig. 6.45(a) considers an open hemisphere radiation and reveals that the higher-order BFD in FDDM provides a bit larger ring radius, and thus, it can resemble a nearly symplectic time integration in smaller N_t and/or larger Δt than those needed for the backward Euler approximation. Fig. 6.45(b) examines a plate scatter and demonstrates that cutting off $\bar{\bar{\mathrm{Z}}}_n$ for $n > N_g$ takes energy out of the system.

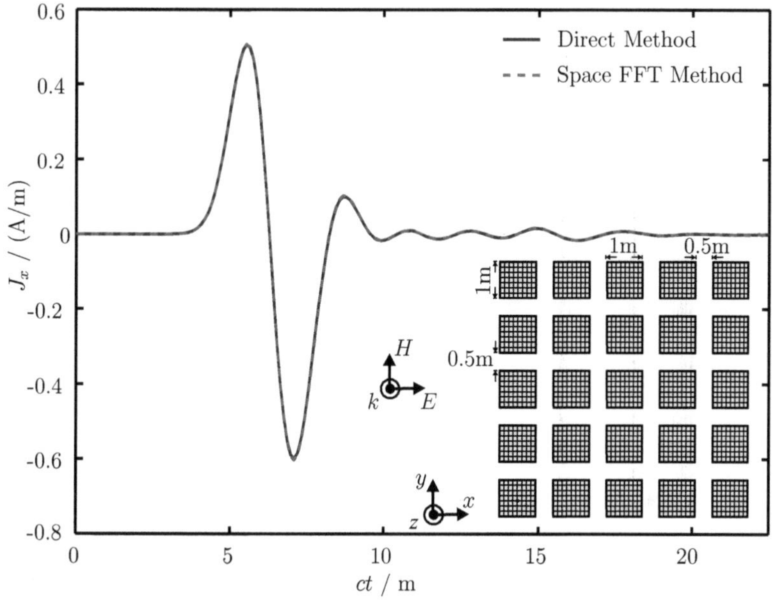

Figure 6.34: Current residing on the middle of the corner cell in the finite FSS consisting of a 5×5 array of plates.

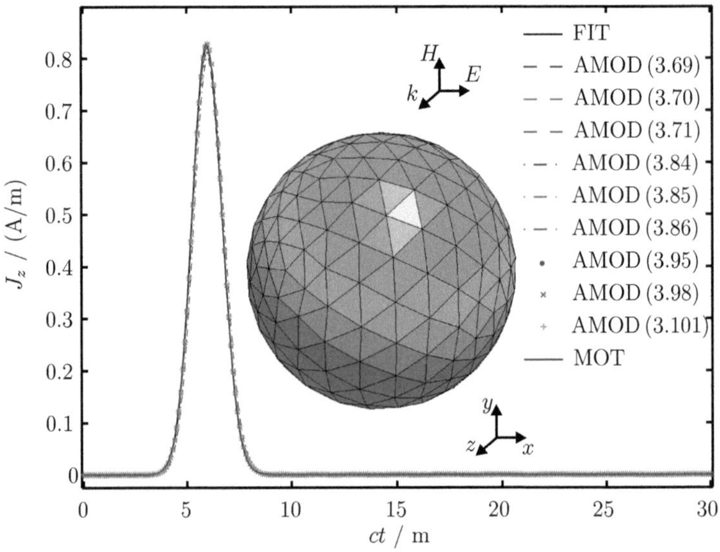

Figure 6.35: z-directed transient induced surface current density on the equator of the conducting sphere; The diverse summation reduction schemes, respectively associated with the matrix equations (3.69), (3.70), (3.71), (3.84), (3.85), (3.86), (3.95), (3.98), and (3.101), preserve the agreement of AMOD solution with the FIT and MOT methods.

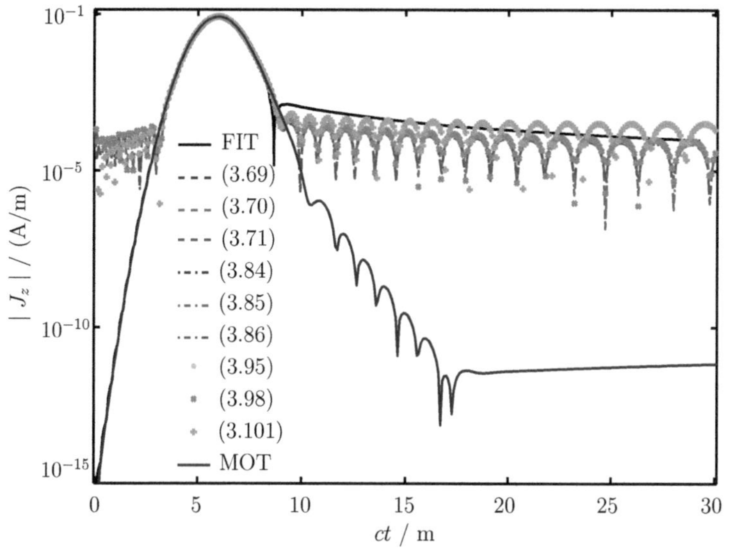

Figure 6.36: Magnitude of the current density specified obtained using different AMOD schemes in Fig. 6.35 in logarithmic scale.

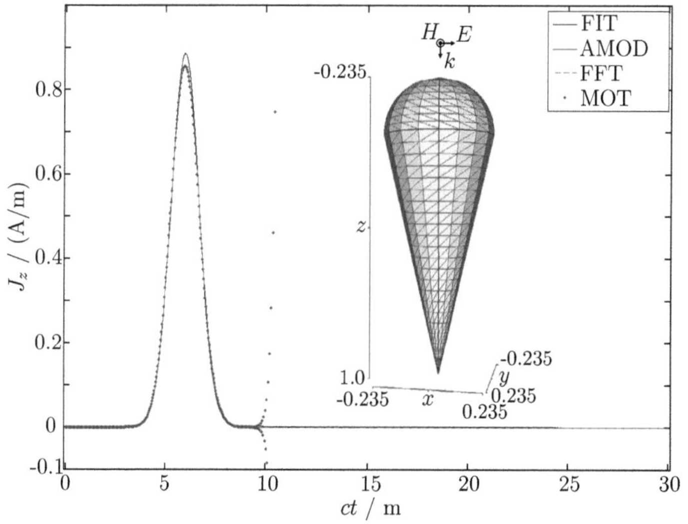

Figure 6.37: Induced current density on the cone sphere at $\mathbf{r} = (0.235, 0, 0)$ under the plane-wave incidence. The FFT implies FFT-besed AMOD solution.

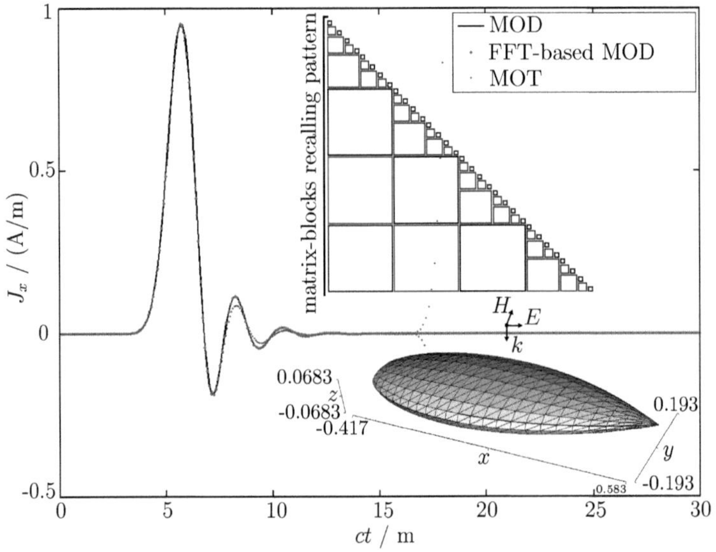

Figure 6.38: The transient induced surface current density on top of the conducting NASA almond calculated by the amalgamation of FFTs on the varying-size and fixed-size aggregates of the retarded interaction matrix blocks.

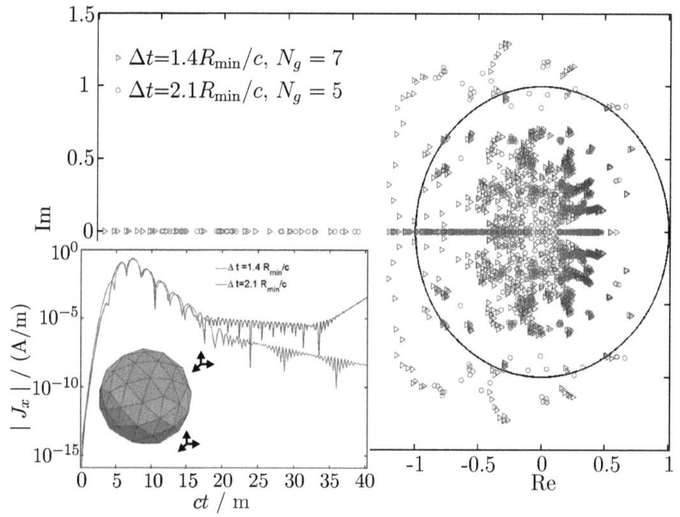

(a) The MOT is generally an energy dissipative method. The numerical approximations in the classical MOT make it an unstable scheme when applied to the (D)EFIE as it always causes at least one eigenvalue outside the unit circle. As long as the hazardous resonance frequencies have not been excited, the response explosion can be postponed using larger time steps.

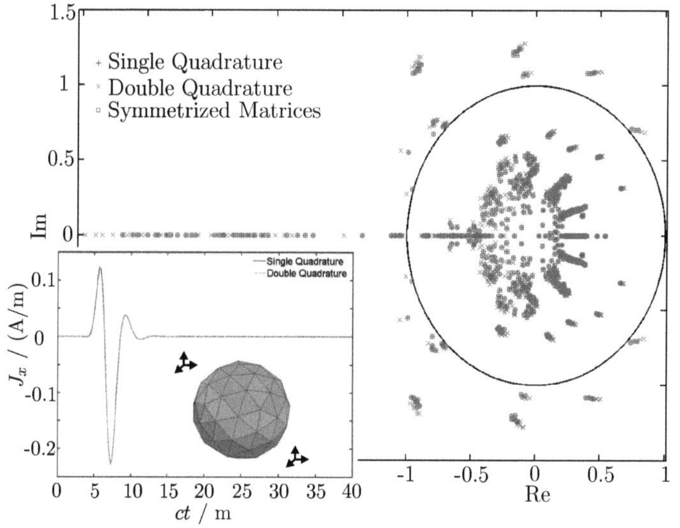

(b) The symmetric quadrature routines alone can not remedy the late-time instability and damping characteristics of the MOT schemes, $\Delta t = 1.75R_{\min}/c$ and $N_g = 7$.

Figure 6.39: The location of MOT system eigenvalues on the complex plane.

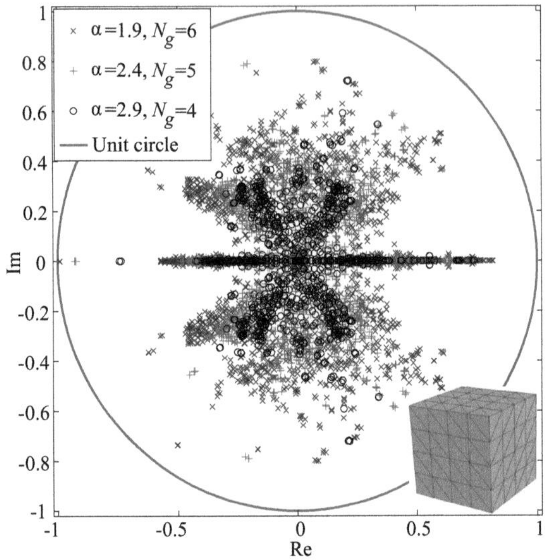

(a) The smaller the time step $\Delta t = \alpha \frac{R_{\min}}{c}$ is chosen (as the more simulation steps N_g takes for the EM fields radiating from source patches to leave the structure), the less dissipation MOT algorithm has.

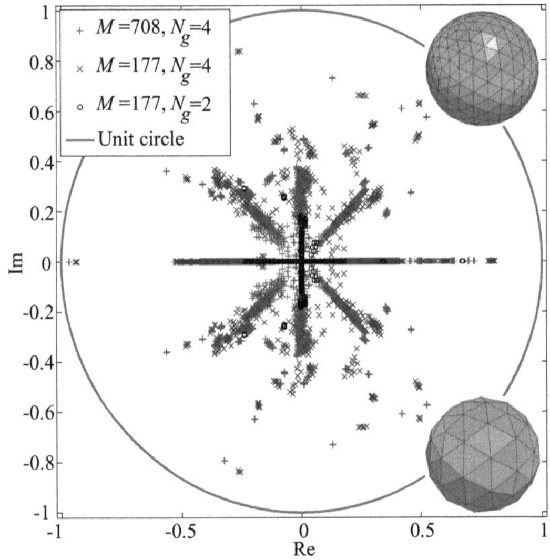

(b) The finer the mesh (the more number of space unknowns M) is used, the MOT dissipates less energy out of the system at the same time step size, $\alpha = 3.8$.

Figure 6.40: The MOT iteration matrix eigenvalues in solving the MFIE for the (a) cube and (b) sphere scatterers. The MOT recipe is stable for the MFIE as all the system eigenvalues lie inside the unit circle.

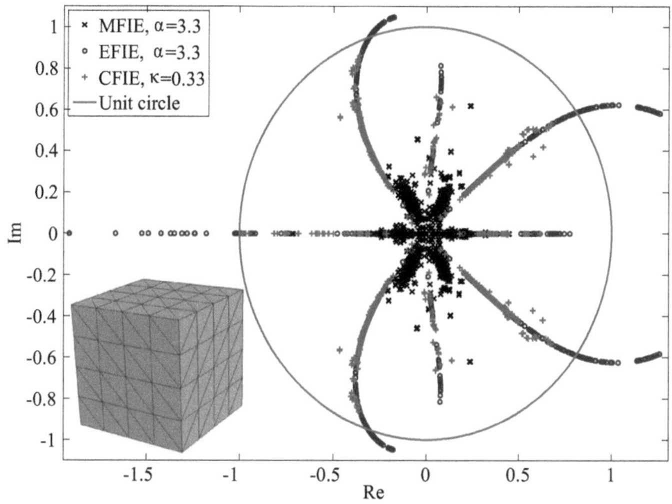

(a) The cube scatterer discretized with $M = 288$ RWG basis functions, N_g=4.

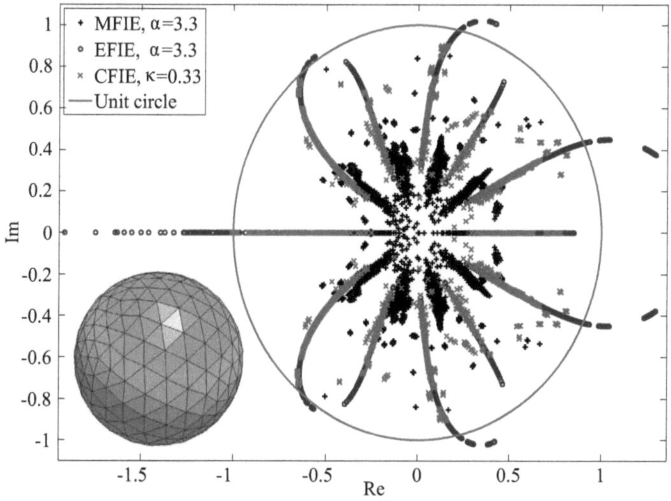

(b) The sphere scatterer discretized with $M = 708$ RWG basis functions, N_g=6.

Figure 6.41: The stabilization effect of the linearly weighted averaging in the CFIE on the dissipative MOT scheme. CFIE=$\frac{\kappa}{\eta}$EFIE+(1-κ)MFIE. N_g is the number of past solution steps contributing in the present solution.

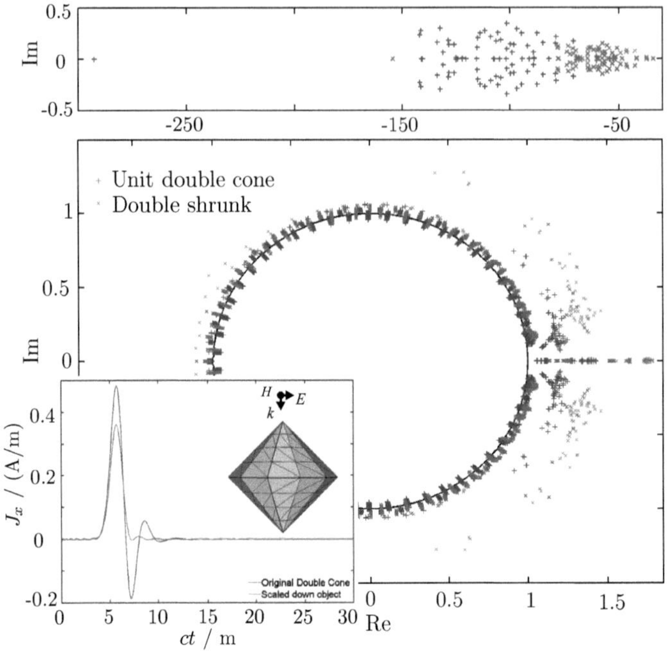

(a) Structure size versus pulse-width ratio

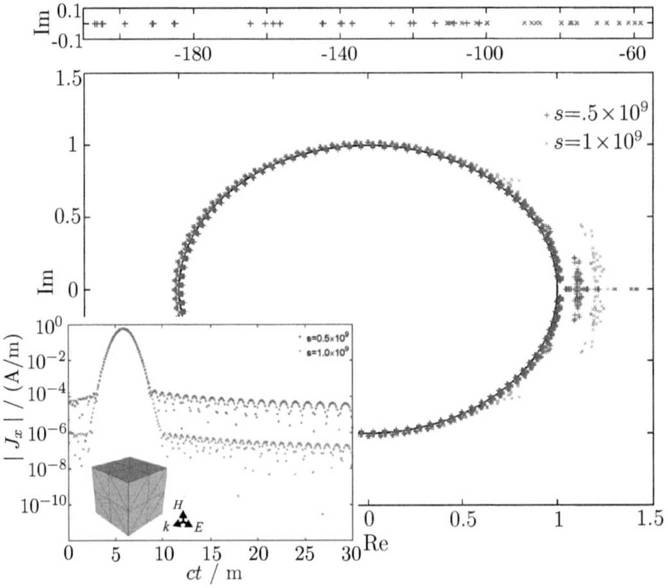

(b) The effect of time resolution scaling

Figure 6.42: The location of MOD system eigenvalues on the complex plane.

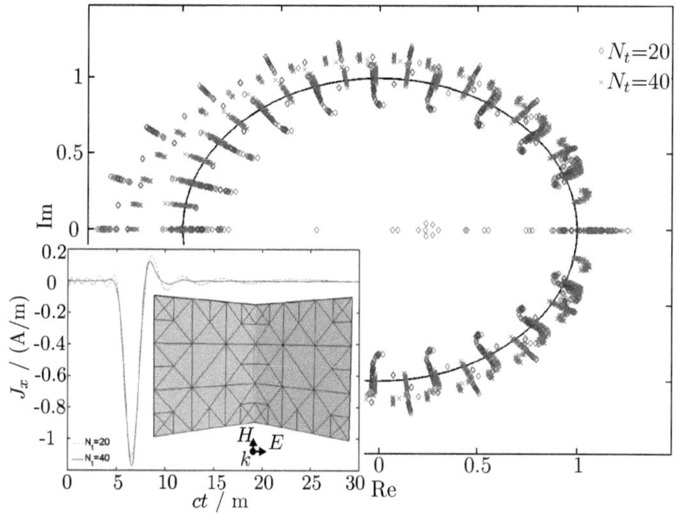

(a) Eigenvalues converge to unit circle as the early ripples are vanished by increasing the number of polynomials

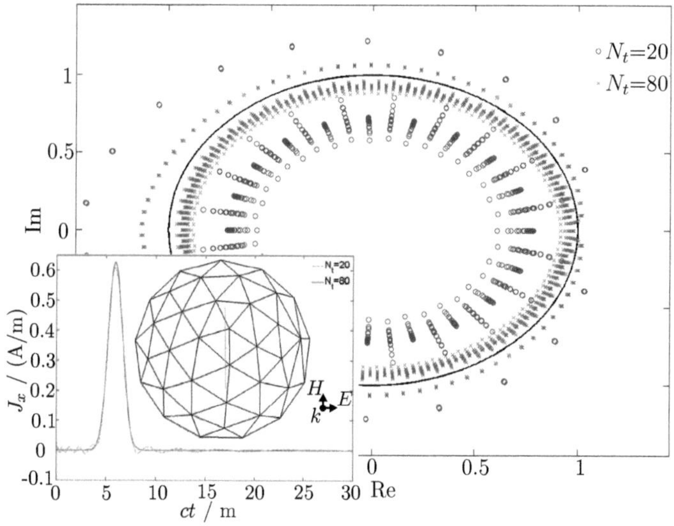

(b) Nearly symplectic method for large N_t

Figure 6.43: The location of AMOD system eigenvalues on the complex plane.

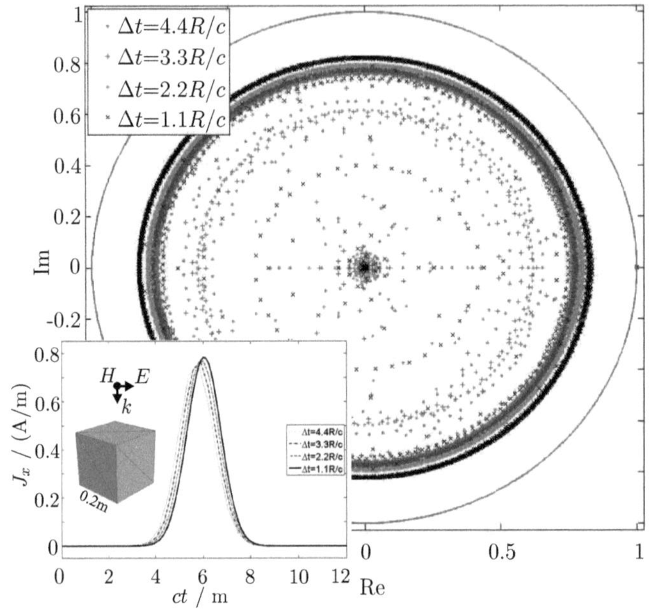

(a) FDDM is nearly energy conservative when time samples are reconstructed closely, $N_t = 250$.

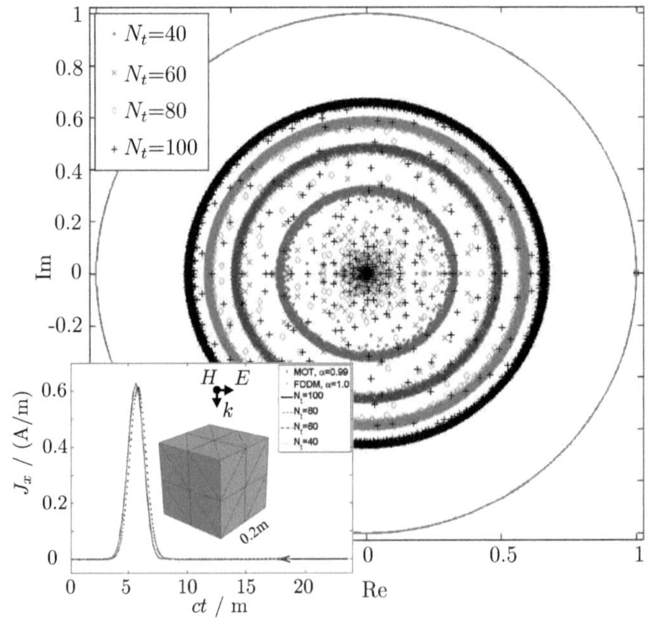

(b) Eigenvalues converge to the unit circle after marching on sufficiently large N_t samples, $\alpha = 5$, $M = 48$.

Figure 6.44: The location of FDDM system eigenvalues on the complex plane.

6.11 Wake Field Simulation in Particle Accelerators

Assuming a point charge Q moving with velocity v along the z-axis of the cylindrical coordinate system close to the speed of light c. The electric field is (Lorentz) contracted into a fictitious thin disk perpendicular to the direction of the charge motion with opening angle proportional to $\sqrt{1-(\frac{v}{c})^2}$. As v approaches c, the radial field lies nearly into the transverse plane and in ultra-relativistic limit, it constricts to a spatial impulsive δ distribution

$$\mathbf{E}^i(\mathbf{r},z,t) = \frac{Q\mathbf{r}}{2\pi\epsilon_0 r^2}\delta(z-ct) \tag{6.3}$$

$$\mathbf{B}^i(\mathbf{r},z,t) = \frac{1}{c}\hat{\mathbf{z}} \times \mathbf{E}^i(\mathbf{r},z,t)$$

where $\mathbf{r}$ is the radial vector. Since the field behind (and ahead) the charged particle Q would be zero, no forces are exerted on a following (or preceding) test particle with charge e.

$$\mathbf{F}^i(\mathbf{r},z,t) = e\left[\mathbf{E}^i(\mathbf{r},z,t) + c\hat{\mathbf{z}} \times \mathbf{B}^i(\mathbf{r},z,t)\right] = 0. \tag{6.4}$$

Therefore, the space charge forces arising from the interaction of particle-to-particle within a charge distribution (with same radial positions) can be neglected for $v \approx c$. The force remains zero when the charges are moving parallel to the z-axis inside a perfectly conducting pipe with arbitrary cross section and translational symmetry along the z-axis; since the field lines terminate on the surface charges that move on the pipe walls synchronously with the original charge(s). However, when the charges are travelling through the linac structures or storage ring vacuum chamber with discontinuities (variation in cross-sectional dimension), wake fields radiate behind the charge as the surface charges confront irregularities along the walls. Due to the (principle of) causality, the wake fields associated with a charge moving along a straight trajectory do not overtake it when $v \approx c$ and never reach to disturb charges travelling ahead. Wake fields interact with the trailing charges moving behind and thus affect the downstream part of the beam.

Similarly in particle accelerators, travelling bunches of charged particles passing through the beam pipes, generate self electromagnetic fields where the magnetic field lines rotate around a fictitious thin disk perpendicular to the direction of particle motion, and the electric field lines transversally terminate on the image surface charges on the wall of the pipe. The self-field of the bunch is the convolution integral of (6.3) for total charge Q with the longitudinal charge distribution $\lambda(z)$

$$\mathbf{E}^i(\mathbf{r},z,t) = \frac{Q\mathbf{r}}{2\pi\epsilon_0 r^2}\lambda(z-ct). \tag{6.5}$$

For the Gaussian bunch with half width (second central moment) σ and normalized to 1,

$$\lambda(s) = \frac{1}{\sqrt{2\pi}\sigma}\exp\left(-\frac{s^2}{2\sigma^2}\right).$$

Thus, the incident radial electric field generated by the charge distribution $Q\lambda(s)$ in the free space

$$\mathbf{E}^i(\mathbf{r},z,t) = \frac{Q\mathbf{r}}{(2\pi)^{\frac{3}{2}}\epsilon_0\sigma r^2}\exp\left(-\frac{(z-ct+ct_0)^2}{2\sigma^2}\right)$$

is applied for the (D)EFIE, and the incident azimuthal ($\hat{\theta}$-directed) magnetic field

$$\mathbf{H}^i(\mathbf{r}, z, t) = \frac{cQ\hat{\mathbf{z}} \times \mathbf{r}}{(2\pi)^{\frac{3}{2}}\sigma r^2} \exp\left(-\frac{(z - ct + ct_0)^2}{2\sigma^2}\right)$$

is considered for exciting the MFIE. The initial position of the bunch can be shifted along the z-axis via the constant ct_0. In cylindrical beam tubes with uniform cross section and PEC walls, as the image charges are moving synchronously with the bunch, regardless of the beam offset to the z-axis, no wake fields are generated. When the bunch arrives at the inhomogeneous accelerator structures like pumps, valves, diagnostics, and other cavity parts with alternating cross sections, however, the induced charges do not synchronously move along the wall, which in turn causes the generation of longitudinal fields so as to keep the boundary condition satisfied over the electrically conductive walls. The wake fields in turn propagate forward and backward and corresponding to the eigen modes of the structure, they oscillate inside the cavity parts for a long time and adversely affect over the dynamics of particles. The so-called long-range wake fields with spectral content below the cut-off frequency of the tube are trapped in the cavity and perturb the motion of other coming charges till the θ_r coupling elements or the ohmic losses of the walls attenuate the resonant modes. Depending on the excitation spectrum, high-frequency parts referred to as the short-range wake fields, are able to follow the bunch through the beam tube and interact with the bunch for a long distance in the tube (known as the interaction region). A catch-up-distance is defined as the distance away from the opening edge or ending tail of discontinuity that the leading charge has to travel so long that it feels the wake field caused by that edge. After the bunch passes the catch-up-distance, accompanying field interaction with the bunch tail gradually reconstructs the energy of the self-field that has been already taken from the bunch by radiation in the cavity. As long as the disc-shaped self-field around the bunch has not been rebuilt, the bunch losses energy and disturbs.[2] Therefore, an in-dept understanding of wake fields is required for proper design of accelerators operating in an optimal status.

Time domain simulation of wake fields has been widely applied using the finite integration technique (FIT) scheme [128]. The particle-in-cell schemes discretizing the whole volume enclosing the model, however, are prone to numerical grid dispersion and the staircase approximation [129]. On the other hand, the time-domain boundary element method only invokes the surface discretization of the structure and more effectively treats the numerical modeling of curved geometries [45, 55, 130, 131, 132] .

Assuming a bunch of particles with the total charge Q moving along the z-axis of the cylindrical coordinate system, the fictitious disk encompassing the self-fields of the beam shrinks to a spacial delta function distribution in the ultra- relativistic limit, exciting an initial incident electric field $\mathbf{E}^i$ (6.5) pointing strictly radially outward from the charges. All the field are zero both ahead and behind the bunch and there are no forces on a test particle either preceding or following this bunch. The beam self-fields illuminate the surrounding structure and induce a corresponding transient electric current $\mathbf{J}$ and the charge density $\nabla \cdot \mathbf{J}$ over the inner surface of the conducting walls. The induced surface current and charges reradiate the scattered fields which can be determined by enforcing the boundary condition on the incident field originated by the presence of the travelling bunch plus the

[2]In the ultra-relativistic case, the motion of the particles is not affected under this rebuilding interactions as the particle energy is infinite.

scattered field over the inside perfectly conducting surface of the vacuum cavity. This leads to the EFIE (2.10), the DEFIE (2.11) or the MFIE (2.24).

6.11.1 Wake Potentials

The influences of wake fields to beam dynamics are of great interest for the design of accelerator components. The concept of wake potential is often used to quantitatively estimate their influences. Considering a charge q is following the exciting bunch Q with the same velocity c along the z-axis, the wake potentials are defined as an indefinite integrals of the normalized wake field forces $\mathbf{F}$ of the bunch over the witness particle at the transverse offset $\mathbf{r} = (x, y)$ along the distance s behind the exciting charge[3] Q. The momentum change for the trailing particle

$$\mathrm{d}\mathbf{p} = \int_{-\infty}^{\infty} \mathbf{F}(\mathbf{r}, z, t)\mathrm{d}t = \int_{-\infty}^{\infty} \mathrm{d}t \left[\mathbf{E}(\mathbf{r}, z, t) + c\,\hat{\mathbf{z}} \times \mathbf{B}(\mathbf{r}, z, t)\right]_{z=ct-s} \tag{6.6}$$

is of interest in longitudinal $\mathrm{d}p_z$ and transverse $\mathrm{d}\mathbf{p}_\perp$ directions. The wake potential is obtained when the momentum change (6.6) is normalized by $\frac{c}{Q}$ factor. Thus, the longitudinal wake potential is expressed as

$$W_\parallel(\mathbf{r}, s) = -\frac{c}{Q}\int_{-\infty}^{\infty} \mathrm{d}t\ E_z(\mathbf{r}, z, t)_{z=ct-s} = -\frac{1}{Q}\int_{-\infty}^{\infty} \mathrm{d}z\ E_z(\mathbf{r}, z, t)_{t=(z+s)/c} \tag{6.7}$$

where the minus sign is to show the energy loss when the leading and trailing charge particles have the same signs, i. e. $W_\parallel < 0$ where the test particle is decelerated and vice versa. The longitudinal wake potential $W_\parallel$ is, in fact, the bunch-induced voltage on the accompanying test particle normalized to the total charge of the excitation bunch (V/C) and it affects the energy of the charge. Similarly, the transverse wake potential, which affects the transverse motion of the bunch, is defined by

$$\begin{aligned}\mathbf{W}_\perp(\mathbf{r}, s) &= \frac{c}{Q}\int_{-\infty}^{\infty} \mathrm{d}t\ \left[\mathbf{E}_\perp(\mathbf{r}, z, t) + c\hat{\mathbf{z}} \times \mathbf{B}(\mathbf{r}, z, t)\right]_{z=ct-s} \\ &= \frac{1}{Q}\int_{-\infty}^{\infty} \mathrm{d}z\ \left[\mathbf{E}_\perp(\mathbf{r}, z, t) + c\hat{\mathbf{z}} \times \mathbf{B}(\mathbf{r}, z, t)\right]_{t=(z+s)/c}.\end{aligned} \tag{6.8}$$

According to the causality principle, the wake fields do not outstrip the bunch and hence for $s < 0$, $W_\parallel \equiv 0$ and $\mathbf{W}_\perp \equiv 0$ where the distance is measured from the head of the bunch ($s = 0$ represents the center for the Gaussian distribution).

When the space-charge Q is travelling on the axis of symmetry of the structure, the Fourier series expansion of the total EM fields yields

$$\begin{aligned}\{E_r, B_\theta, E_z\}(r, \theta, z, t) &= \sum_{m=0}^{\infty}\{e_r, b_\theta, e_z\}^{(m)}(r, z, t)\cos m\theta \\ \{B_r, E_\theta, B_z\}(r, \theta, z, t) &= \sum_{m=0}^{\infty}\{b_r, e_\theta, b_z\}^{(m)}(r, z, t)\sin m\theta\end{aligned} \tag{6.9}$$

[3]In practice, particles move slower than the speed of light and the fields forereach the particles. The overtaking fields, however, are diminished by the wall dissipation and/or obstacles in the pipe.

where m is the azimuthal mode number. The field components associated with $m = 0$ are called the monopole mode, $m = 1$ dipole mode, $m = 2$ quadruple mode, and so on. The monopole mode field corresponds to the axially symmetric field. For the axially symmetric field only E_r, B_θ, and E_z components are nonzero. As the total scattered (radiated) field satisfies the source-free homogeneous Maxwell equations (2.1), the azimuthal component of the Faraday's law gives

$$\left(\nabla \times \mathbf{E} = -\frac{\partial \mathbf{B}}{\partial t}\right)_\theta \longrightarrow \frac{\partial E_z}{\partial r} = \frac{\partial E_r}{\partial z} + \frac{\partial B_\theta}{\partial t}. \tag{6.10}$$

The partial derivatives can be converted to the wake potential integration variable dz in (6.7) for any generic field D using

$$\frac{\mathrm{d}}{\mathrm{d}z} D(z,s) = \left(\frac{\partial}{\partial z} + \frac{\partial}{c\partial t}\right) D(z, t(z,s))\bigg|_{t=\frac{z+s}{c}}. \tag{6.11}$$

Therefore,

$$(6.10) \longrightarrow \frac{\partial}{\partial r} E_z = \frac{\mathrm{d}}{\mathrm{d}z}(E_r + cB_\theta) - \frac{\partial E_r}{c\partial t} - c\frac{\partial B_\theta}{\partial z}. \tag{6.12}$$

Analogously, the radial component of the Ampere's law gives

$$\left(\nabla \times \mathbf{B} = \frac{\partial \mathbf{E}}{c^2 \partial t}\right)_r \longrightarrow \frac{\partial E_r}{c^2 \partial t} = \frac{\partial B_z}{r\partial \theta} - \frac{\partial B_\theta}{\partial z}. \tag{6.13}$$

Considering the rotational symmetric field assumption ($\partial_\theta = 0$) for (6.13) and then substituting it in (6.12), the relation is simplified to

$$\frac{\partial}{\partial r} E_z = \frac{\mathrm{d}}{\mathrm{d}z}(E_r + cB_\theta). \tag{6.14}$$

Integrating (6.14) along the z coordinate, the wake potential definition (6.7) results in

$$\frac{\partial}{\partial r} W_\parallel = -\frac{1}{Q}[E_r + cB_\theta]_{-\infty}^{+\infty} = 0 \tag{6.15}$$

as all scattered fields vanish at infinity. Therefore, the path of the integration in the wake potential calculations does not depend on parallel displacement(s) as long as it remains inside the beam tube. The independency of integration paths in wake potential calculations allows one to integrate along the beam pipe surface where $E_z = 0$ and its contribution vanishes.

The two Maxwell's equations contains other components:

$$\left(\nabla \times \mathbf{E} = -\frac{\partial \mathbf{B}}{\partial t}\right)_r \longrightarrow \frac{\partial E_z}{r\partial \theta} - \frac{\partial E_\theta}{\partial z} = -\frac{\partial B_r}{\partial t} \tag{6.16}$$

$$\left(\nabla \times \mathbf{B} = \frac{\partial \mathbf{E}}{c^2 \partial t}\right)_\theta \longrightarrow \frac{\partial E_\theta}{c^2 \partial t} = \frac{\partial B_r}{\partial z} - \frac{\partial B_z}{\partial r}. \tag{6.17}$$

Combining (6.10) with (6.13) and (6.16) with (6.17), after (6.11) is applied to them, can be written respectively as:

$$\frac{\partial}{\partial r}E_z - \frac{\partial}{\partial z}(E_r + cB_\theta) = -\frac{1}{r}\frac{\partial}{\partial \theta}cB_z$$

$$\frac{\partial}{\partial r}cB_z + \frac{\partial}{\partial z}(E_\theta - cB_r) = \frac{1}{r}\frac{\partial}{\partial \theta}E_z$$

The multiple expansion of the scattered fields as (6.9) yields

$$\left[\frac{\partial}{\partial r}e_z - \frac{\partial}{\partial z}\left(e_r + cb_\theta\right)\right]^{(m)} = -\frac{m}{r}[cb_z]^{(m)}$$

$$\left[\frac{\partial}{\partial r}cb_z + \frac{\partial}{\partial z}\left(e_\theta - cb_r\right)\right]^{(m)} = -\frac{m}{r}[e_z]^{(m)}$$

for each order m. The sum and difference of the above two equations gives

$$\frac{\partial}{\partial r}\left(r^m\left[e_z + cb_z\right]^{(m)}\right) = \frac{\partial}{\partial z}\left(r^m\left[e_r + cb_\theta - e_\theta + cb_r\right]^{(m)}\right) \tag{6.18}$$

$$\frac{\partial}{\partial r}\left(r^{-m}\left[e_z - cb_z\right]^{(m)}\right) = \frac{\partial}{\partial z}\left(r^{-m}\left[e_r + cb_\theta + e_\theta - cb_r\right]^{(m)}\right). \tag{6.19}$$

As a result, the two vector components that can be defined by the parentheses (gradients of potential) are closed (irrotational) in (r, z) plane, i.e.,

$$\frac{\partial}{\partial r}S_z^{(m)}(r,z,s) - \frac{\partial}{\partial z}S_r^{(m)}(r,z,s) = 0$$

$$\frac{\partial}{\partial r}D_z^{(m)}(r,z,s) - \frac{\partial}{\partial z}D_r^{(m)}(r,z,s) = 0.$$

This implies that the vector integrals along any closed contour (arbitrarily path enclosing the vacuum) vanish, and thus, the wake field infinite integration path along a straight line contour can be replaced by any finite arbitrary contour C spanning the structure longitudinally.

In boundary integral equation approach it would be more convenient to choose a contour across the structure which coincide with the boundary walls, ending on the beam tube with identical end radii. As a result, in the case of monopole mode, the longitudinal wake potential is r-independent and is reduced to

$$\begin{aligned} W_{\|}(\mathbf{r},s) &= -\frac{1}{Q}\int_C \mathbf{S}^{(m=0)}(\mathbf{r},z,s)\cdot \mathrm{d}\mathbf{l} \\ &= -\frac{1}{Q}\int_C [e_z \mathrm{d}z + (e_r + cb_\theta)\mathrm{d}r](\mathbf{r},z,s) \\ &= -\frac{1}{Q}\int_C [\mathbf{E}_l\cdot \mathrm{d}\mathbf{l} + cB_\theta \mathrm{d}r](\mathbf{r},z,t(z,s)). \end{aligned} \tag{6.20}$$

Note that since the radii of the beam tubes are identical on both sides of the structure ($r_{in} = r_{out}$), the additional logarithm term $\frac{1}{\pi\epsilon_0}\ln\left[\frac{r_{\text{out}}}{r_{\text{in}}}\right]\lambda(s)$ is vanished [133]. The integration over the tangential component of the induced electric field $\mathbf{E}_l\cdot\mathrm{d}\mathbf{l} = E_z\mathrm{d}z + E_r\mathrm{d}r$ is terminated as well, when the whole path of integration passes over the perfect conducting wall. To incorporate the evaluated surface current density with (6.20), the azimuthal component of the magnetic flux density can be substituted by

$$\mathbf{n}\times\mathbf{B} = \mu\mathbf{J}$$

and thus, we have

$$W_{\|}(\mathbf{r},s) = -\frac{\mu c}{Q}\int_C J_l(r,\theta,z,\frac{z+s}{c})\mathrm{d}r. \tag{6.21}$$

Since the tube radius does not changes on the cylindrical parts, $\mathbf{J}\cdot\mathrm{d}\mathbf{r} = 0$ and the beam pipe contributions vanish. Thus, the surface current density has to be integrated only along the axial parts of the contour. The contour can be chosen along the intersection of the boundary wall and a cutting half plane, e.g. $x = 0$, $y > 0$ ($\theta = \frac{\pi}{2}$). When a typical triangle T_k^q with vertices $\mathbf{r}_i \in (\mathbf{r}_1,\mathbf{r}_2,\mathbf{r}_3)$ intersects with the cutting half plane in points $(\mathbf{r}_0,\mathbf{r}_0')$, one can define a local coordinate $0 \le \eta \le 1$ for any arbitrary point $\mathbf{r}$ along the longitudinal path $\mathbf{l}$ as

$$\begin{aligned}\mathbf{l}(\eta) &= \mathbf{r}_0 + (\mathbf{r}_0' - \mathbf{r}_0)\eta\\ \mathrm{d}\mathbf{l}(\eta) &= (\mathbf{r}_0' - \mathbf{r}_0)d\eta.\end{aligned}$$

Using the RWG BF, the spatial variation of the current along η is represented by

$$\sum_i q_i l_k^i I_k^i \rho_i = \sum_i q_i l_k^i I_k^i\left[(\mathbf{r}_0 - \mathbf{r}_i) + (\mathbf{r}_0' - \mathbf{r}_0)\eta\right].$$

multiplied by $\frac{T(\frac{z+s}{c})}{2A_k}$. Therefore, the contribution within every triangle can be calculated analytically through

$$\begin{aligned}\sum_i q_i l_k^i I_k^i\int_{C\bigcap T_k^q}\rho_i\cdot\mathrm{d}\mathbf{r} &= \sum_i q_i l_k^i I_k^i\oint_0^1\left[(\mathbf{r}_0 - \mathbf{r}_i) + (\mathbf{r}_0' - \mathbf{r}_0)\eta\right]\cdot(\mathbf{r}_0' - \mathbf{r}_0)\mathrm{d}\eta\\ &= \sum_i q_i l_k^i I_k^i\cos(\varphi)\left[(y_0 - y_i) + (y_0' - y_0)/2\right](y_0' - y_0)\end{aligned} \tag{6.22}$$

where $\oint$ implies the integration on the radial component (truncation of $\hat{\mathbf{z}}$-directed contribution) and $\cos(\varphi) = \frac{(\mathbf{r}_0'-\mathbf{r}_0)\cdot\mathbf{a_r}}{|\mathbf{r}_0'-\mathbf{r}_0|}$, $\mathbf{a_r} = \hat{\mathbf{y}}$. The longitudinal component suffices to reconstruct the transverse components of the wake potential using Maxwell's equation. The relation between the longitudinal and transverse wake potentials is known as the Panofsky-Wenzel theorem

$$\partial_s W_\perp(\mathbf{r},s) = \nabla_\perp W_{\|}(\mathbf{r},s). \tag{6.23}$$

Thus, the integration of the transverse gradient of the longitudinal wake potential gives the transverse ones

$$W_\perp(\mathbf{r},s) = \nabla_\perp\int_{-\infty}^{s} W_{\|}(\mathbf{r},s')\mathrm{d}s'. \tag{6.24}$$

The longitudinal wake potential is a harmonic function of the transverse coordinates, $\nabla_\perp^2 W_{\|}(\mathbf{r},s) = 0$. This property can be used to efficiently evaluate the radial dependency of the wake potential in cylindrical symmetric structures. Describing the beam as a current source, a lumped circuit can be also used to model the beam-surrounding interactions by the voltage per unit charge that the wake potential put on an impedance. The impedance would be the Fourier transform of the wake field on the co-moving frame $s = ct - z$

$$\mathbf{Z}(\frac{\omega}{c}) = \int_{-\infty}^{\infty} \mathbf{W}(s) \exp(-j\omega s) \mathrm{d}s. \tag{6.25}$$

In contrary to the frequency-domain codes that are suitable for long-range wake field calculation, the time-domain solvers are of more interest for long-time propagation of the short-range (high-frequency) wake field [134, 135].

6.11.2 Cylindrical, Pillbox, and Tesla Cell Cavities

For investigation of wake-fields excited by moving charged particles in accelerator cavities, it is usually supposed that tube-arm waveguides are infinite along the z-axis. The effect of infinitely long pipes, however, can not be properly modeled by the bounded 3D mesh. An air-filled round cylindrical waveguide with PEC sidewalls [136] is considered first to study the excitation of electromagnetic oscillations by electron bunches. As shown in Fig. 6.46, the effect of confined mesh in the BEM does not let modeling a bunch to be entered into the structure without generating wake-fields. To avoid distortion of the field distribution when the bunch is entering to or leaving infinitely long tubes, an absorbing boundary condition (ABC) is devised for the open ending patches in the BEM by defining extended fictitious edge elements (additional degrees of freedom) on the cross sections of the meshed tube arms according to Fig. 6.47. Incorporating the ABC into the open mesh ends and using the EFIE formulation [137], charged particles can enter into the beam tube without arising any behind fields and they can leave the beam tube with out any reflection, Fig. 6.48.

A closed cylindrical (pillbox with tube arms) cavity, 0.1 m-long with tube radius $r_{\text{pill}} = 0.04$ m and additionally attached beam tubes of radius $r_{\text{tube}} = 0.02$ m as depicted in Fig. 6.50, is subdivided monotonically into 4842 triangular patches respectively with 7263 common edges. A Gaussian-shaped bunch of charged particles with $Q = 1$ nC and the standard deviation of $c\sigma = 0.025$ m passes along the z-axis with the velocity of light. The self-field of the beam forms an initial incident electric field $\mathbf{E}^i$ in (6.3). After applying the moment expansion and the Galerkin's testing procedure, (3.19) is constructed for the axisymmetric 3D model and the resultant matrix equation is solved for the time step size of $\Delta t = \alpha \frac{R_{min}}{c}$ with $\alpha = 2.6$. Fig. 6.49 shows the longitudinal component of the induced surface current (total tangential field) in the middle of the pillbox cavity obtained by the MOT and the CST PARTICLE STUDIO® simulation package, version 2009 [56] run with 204,800 cells. Note that the modal analysis of the cavity when the bunch leaves the structure confirms the oscillation amplitude accuracy of the CST results. It seems that after the wake fields reach to the ends of the confined model, reflections perturb the boundary element results whereas the absorbing boundary conditions in the FIT let the infinitely long tube arms be incorporated to the model. As introduced by Napoly [138], one can derive from the potential in (r, z) plane, two-dimensional vectors whose integral along a closed contour enclosing the vacuum vanishes, and thus, deform the wake field integration path from a straight line at constant radius for (6.7), to any contour across the structure. The indirect integration method for the wake potential calculation in the axisymmetric structures (6.7) is applied. The wake potential have been shown in Fig. 6.50

using the FIT results and those obtained using the EFIE based method together with Crank-Nicolson integrator. Fig. 6.51 shows the surface current distribution (image charge movements) on the pillbox cavity at different time instances $t_n = n\Delta t$ when the bunch is passing through the axis of the structure. Similarly, Fig. 6.52 illustrates the surface current distribution (image charge movements) on a TESLA single cell cavity at different time instances $t_n = n\Delta t$ when the bunch is passing through the axis of the structure.

6.12 Realistic Complex Structures

In this section the applicability of the MOT scheme to large scale EM scattering problems is investigated. The parallel direct sparse solver (Pardiso) in Intel Math Kernel Library (MKL) is used for multiple solving of the large matrix equations. First, a 14.6 m-long Glider (13.75 m distance between the end of wings) is illuminated by a z-polarized Gaussian plane-wave traveling along $\hat{\mathbf{k}} = \hat{\mathbf{x}}$ with $cT = 20$ m, $ct_0 = 30$ m. The glider is meshed by 7314 triangles and the EFIE is solved for $M = 10971$ degrees of freedoms using the MOT with safety factor $\alpha = 35$. Fig. 6.53 demonstrates the induced surface current magnitude on the glider at different time instances $t_n = n\Delta t$ where $c\Delta t = 1.197$ m. As the next large scale problem, a 130.8 m-long ship is illuminated by a z-polarized Gaussian plane-wave traveling along $\hat{\mathbf{k}} = \hat{\mathbf{x}}$ with $cT = 40$ m, $ct_0 = 120$ m. The ship is meshed by 10472 triangles and the MFIE is solved for $M = 15708$ degrees of freedoms using the MOT with safety factor $\alpha = 15$. Fig. 6.54 shows the induced surface current magnitude on the ship at different time instances $t_n = n\Delta t$ where $c\Delta t = 2.996738$ m. For $n \leq 120$ the solution procedure of the both examples take less than half a day on a single 64-bit machine. The exact size and details of the both glider and ship geometries can be found in CST MICROWAVE STUDIO® 2008 examples. When the ship case study is solved for $M = 32925$ RWG elements, the simulation run time takes about two days. As the last example an Airbus A380 model (123.1915 m-long with 110.7 m distance between the end of wings) is illuminated by a y-polarized Gaussian plane-wave traveling along $\hat{\mathbf{k}} = -\hat{\mathbf{z}}$ with the pulse specifications $cT = 80$ m, $ct_0 = 120$ m. The airplane surface is meshed by 13032 triangles and the CFIE is solved for $M = 19548$ degrees of freedoms using the MOT with safety factor $\alpha = 5 \times 10^5$. Fig. 6.55 illustrates the induced surface current magnitude on the airplane at different time instances $t_n = n\Delta t$ when $c\Delta t = 3.620804$ m.

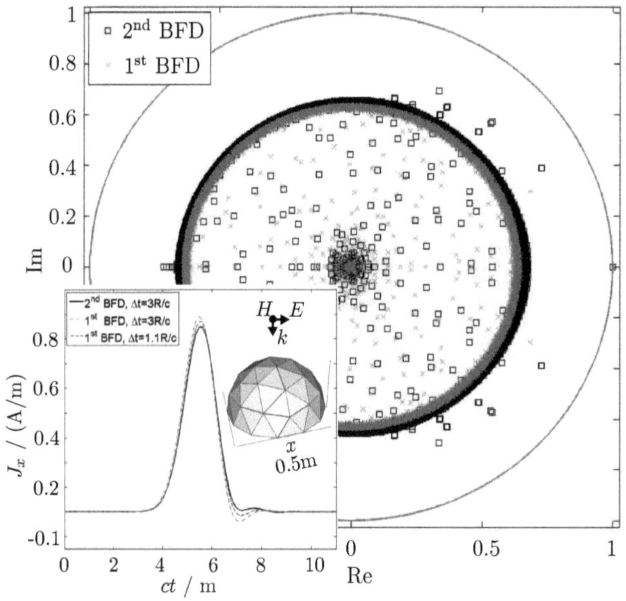

(a) 1st order BFD solution converges to the 2nd one for small time steps.

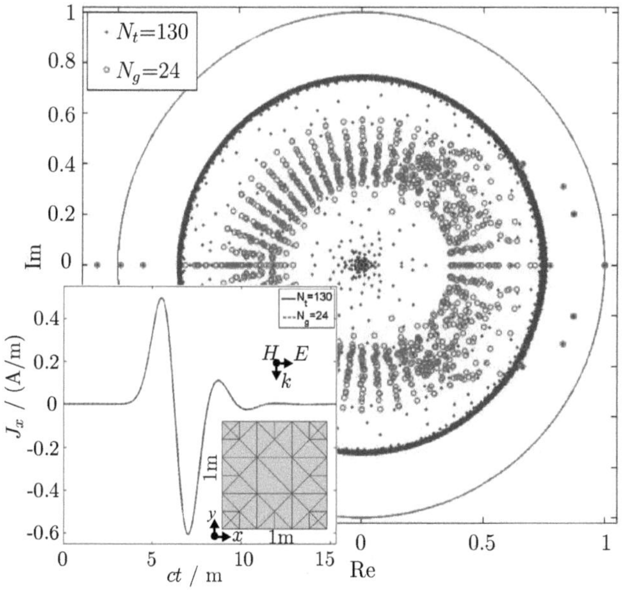

(b) Cut-off strategy leads to energy dissipation.

Figure 6.45: The location of FDDM system eigenvalues on the complex plane.

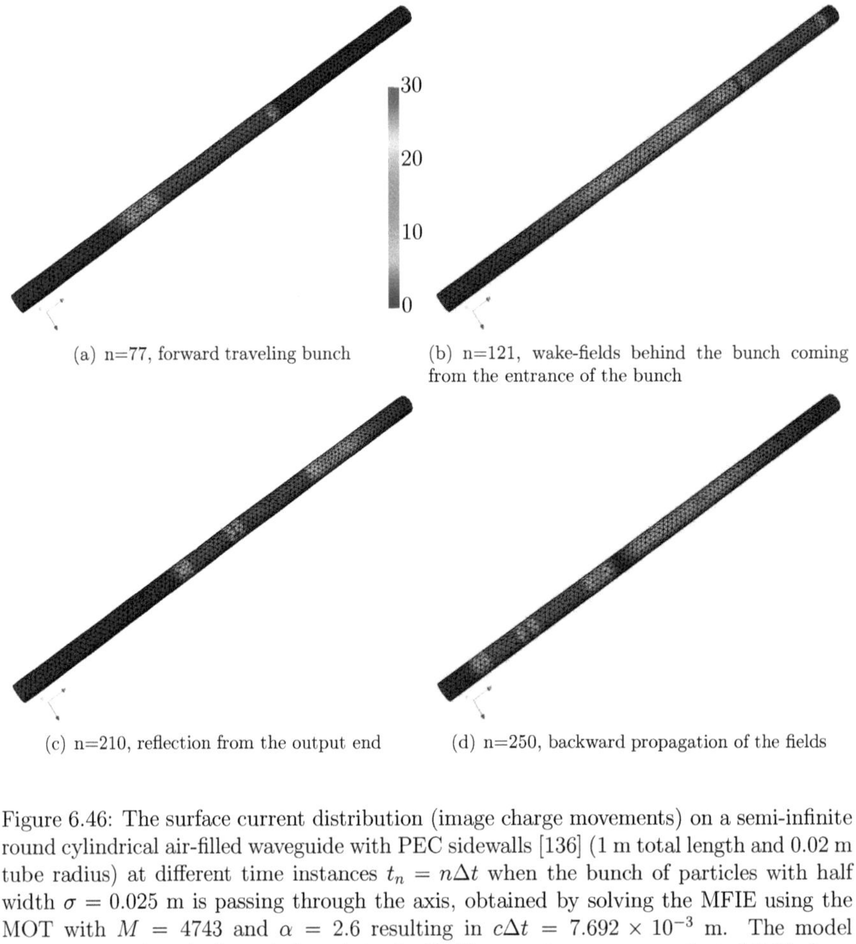

(a) n=77, forward traveling bunch

(b) n=121, wake-fields behind the bunch coming from the entrance of the bunch

(c) n=210, reflection from the output end

(d) n=250, backward propagation of the fields

Figure 6.46: The surface current distribution (image charge movements) on a semi-infinite round cylindrical air-filled waveguide with PEC sidewalls [136] (1 m total length and 0.02 m tube radius) at different time instances $t_n = n\Delta t$ when the bunch of particles with half width $\sigma = 0.025$ m is passing through the axis, obtained by solving the MFIE using the MOT with $M = 4743$ and $\alpha = 2.6$ resulting in $c\Delta t = 7.692 \times 10^{-3}$ m. The model demonstrates that the bounded mesh in the BEM not only generates reflected fields from the both endings rather it causes non-physical wake-fields behind the bunch.

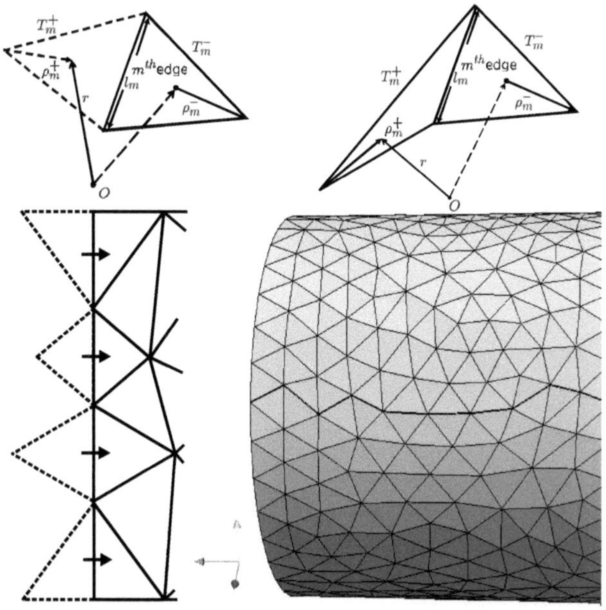

Figure 6.47: Definition of additional current elements on all open patches to properly let the incoming and outgoing self-field of the bunch be modeled.

(a) n=40, forward traveling bunch

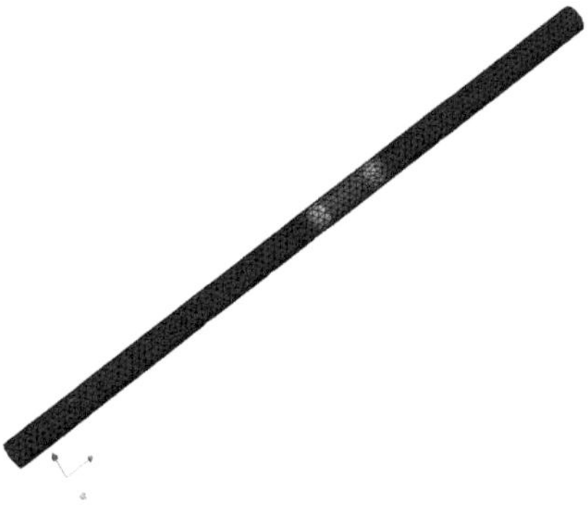

(b) n=90, no fields behind the bunch

(c) n=135, fields right before the bunch leaves the tube

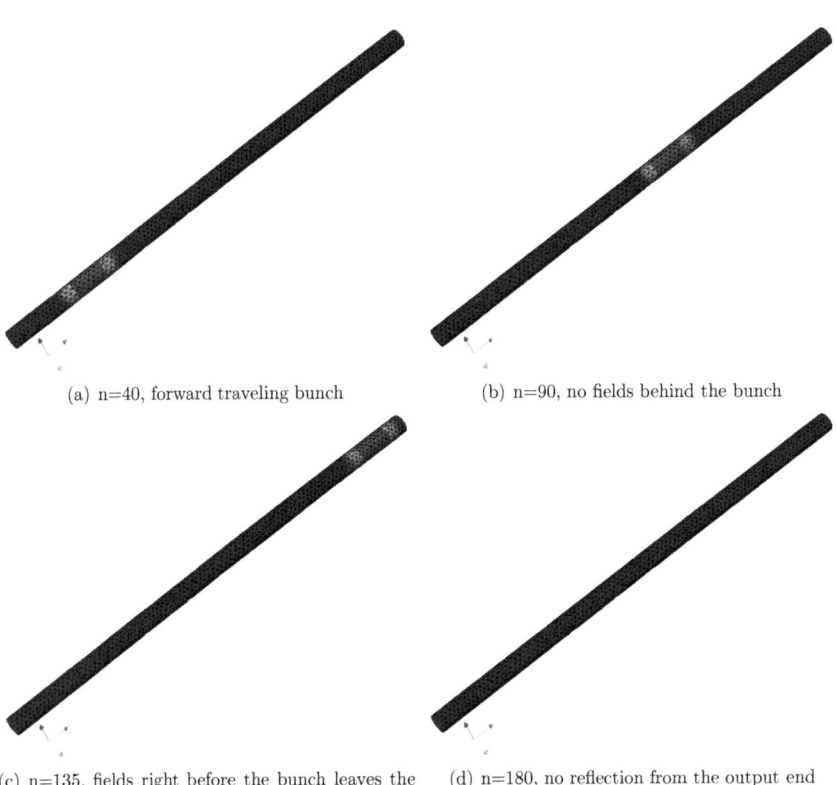

(d) n=180, no reflection from the output end

Figure 6.48: The surface current distribution (image charge movements) on a semi-infinite round cylindrical air-filled waveguide with PEC sidewalls (1 m total length and 0.02 m tube radius) at different time instances $t_n = n\Delta t$ when the bunch of particles with half width $\sigma = 0.025$ m is passing through the axis, obtained by solving the EFIE using the MOT equipped with the ABC for $M = 4743$ and $\alpha = 2.6$ resulting in $c\Delta t = 7.692 \times 10^{-3}$ m.

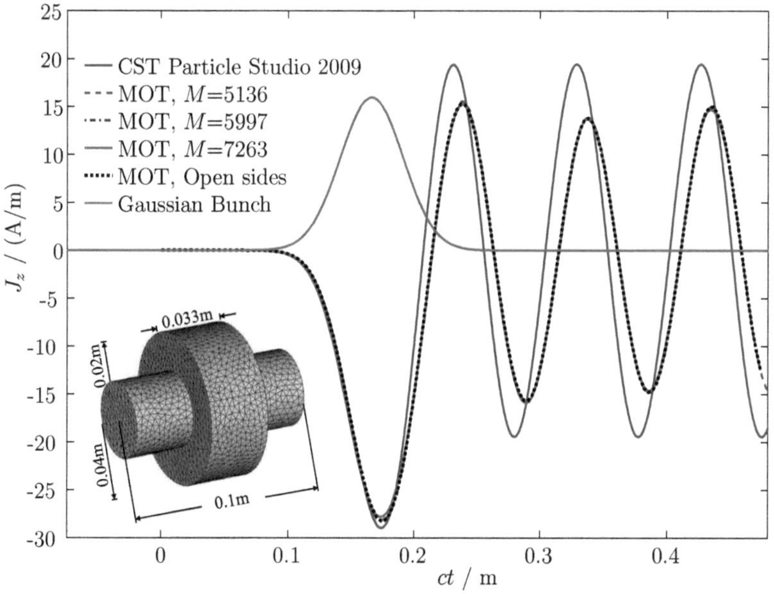

Figure 6.49: The z-directed induced surface current density at the middle of a closed pillbox cavity when a bunch of electrons with total charge $Q = -10^{-9}$ C and half-width $\sigma = 0.025$ m passes through the axis of the structure. $\alpha = 2.6$ and M indicates the unknown edge elements in solving the TDIE. The other two components of the current have negligible values.

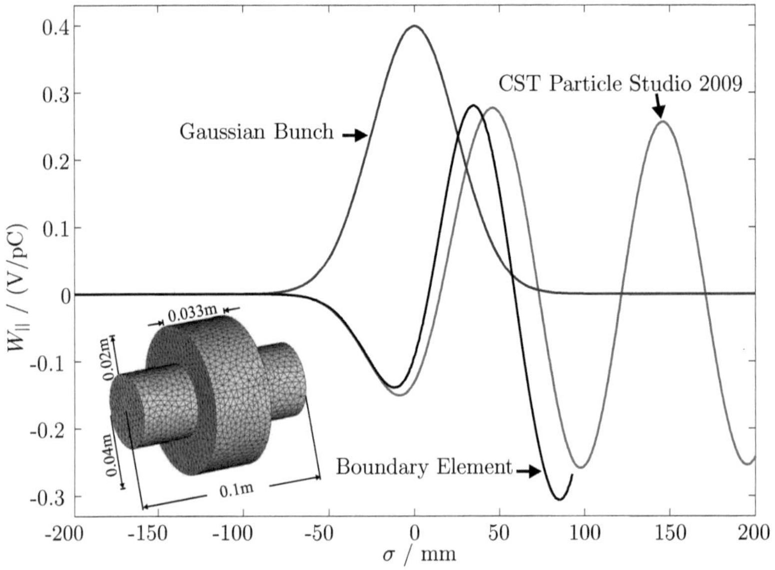

Figure 6.50: The longitudinal wake potential of the pillbox cavity.

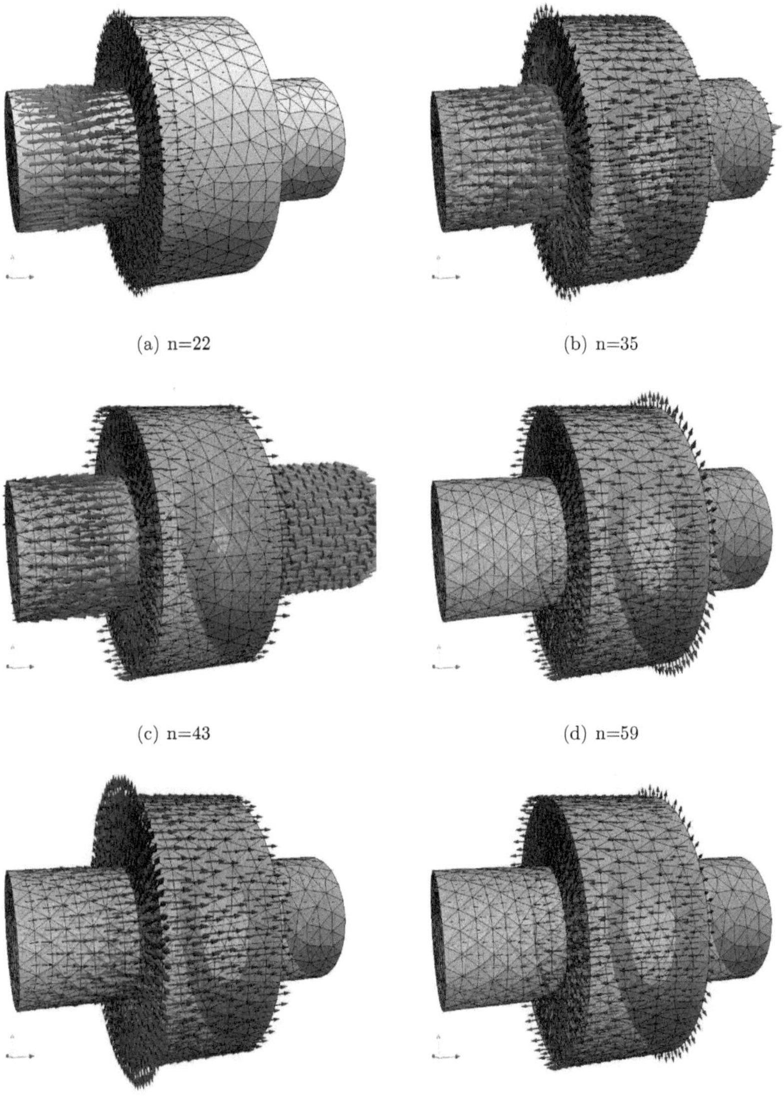

(a) n=22 (b) n=35

(c) n=43 (d) n=59

(e) n=69 (f) n=78

Figure 6.51: The surface current distribution (image charge movements) on a pillbox cavity ($L = 0.1$ m, $r_{\text{tube}} = 0.02$ m, $r_{\text{pill}} = 0.04$ m) at different time instances $t_n = n\Delta t$ when the bunch of particles with half width $\sigma = 0.025$ m is passing through the axis, $\alpha = 2.6$ resulting in $c\Delta t = 4.9321 \times 10^{-3}$ m.

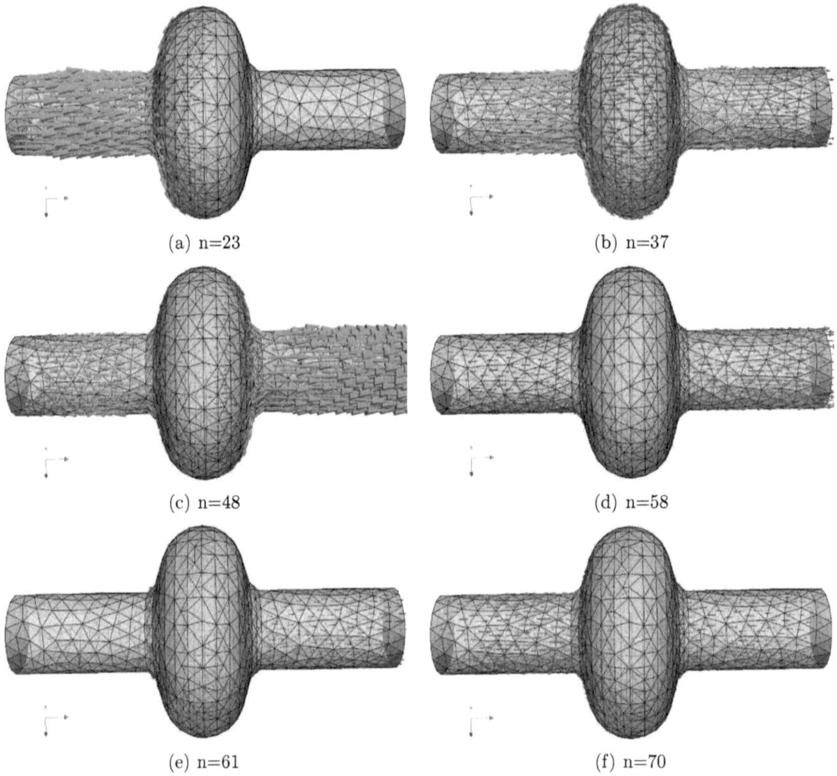

Figure 6.52: The surface current distribution (image charge movements) on the first asymmetric cell of a Tesla cavity ($L = 0.383$ m, $r_{\text{tube}} = 0.039$ m, $r_{\max} = 0.1033$ m) at different time instances $t_n = n\Delta t$ when the bunch of particles with half width $\sigma = 0.1$ m is passing through the axis, $\alpha = 3.9$ resulting in $c\Delta t = 1.524 \times 10^{-2}$ m.

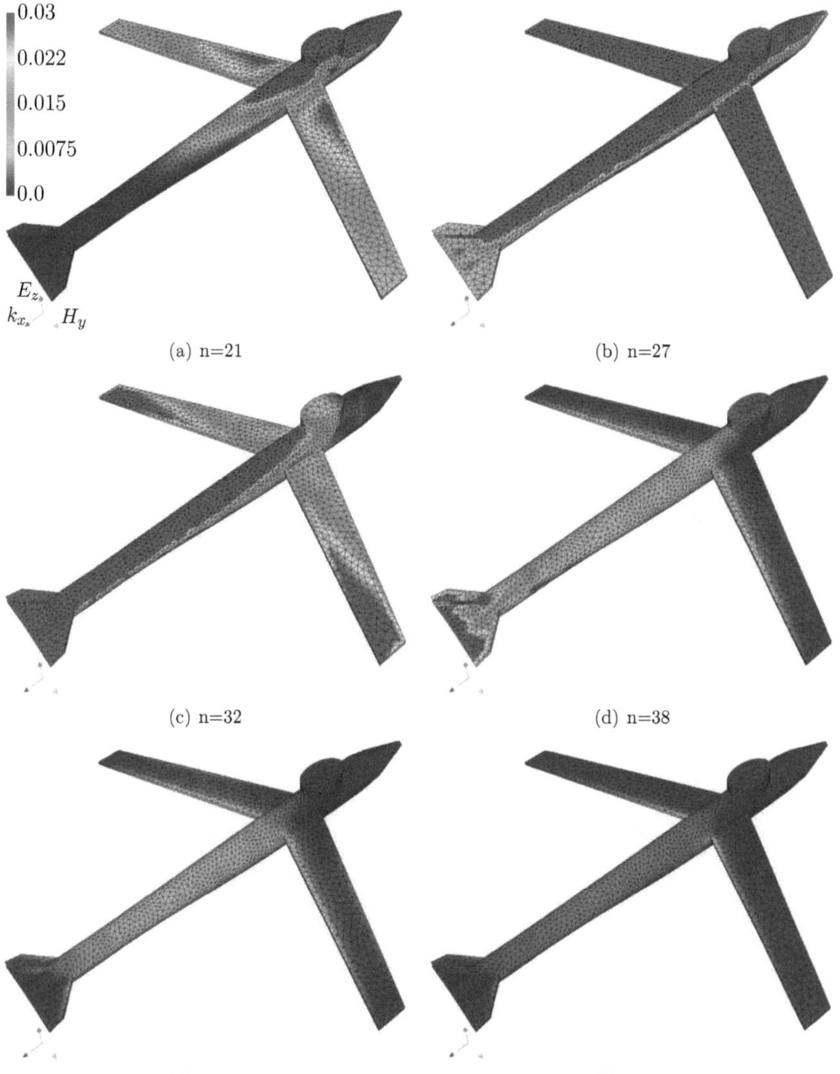

(a) n=21 (b) n=27

(c) n=32 (d) n=38

(e) n=49 (f) n=65

Figure 6.53: The induced surface current magnitude on a 14.6 m-long glider (13.75 m distance between the end of wings) at different time instances $t_n = n\Delta t$ illuminated by a z-polarized Gaussian plane-wave traveling along $\hat{\mathbf{k}} = \hat{\mathbf{x}}$ with the pulse specifications $cT = 20$ m, $ct_0 = 30$ m. The glider is meshed by 7314 triangles and the EFIE is solved for $M = 10971$ degrees of freedoms using the MOT with safety factor $\alpha = 35$ resulting in $c\Delta t = 1.197$ m. The geometry can be found in CST MICROWAVE STUDIO® 2008 examples.

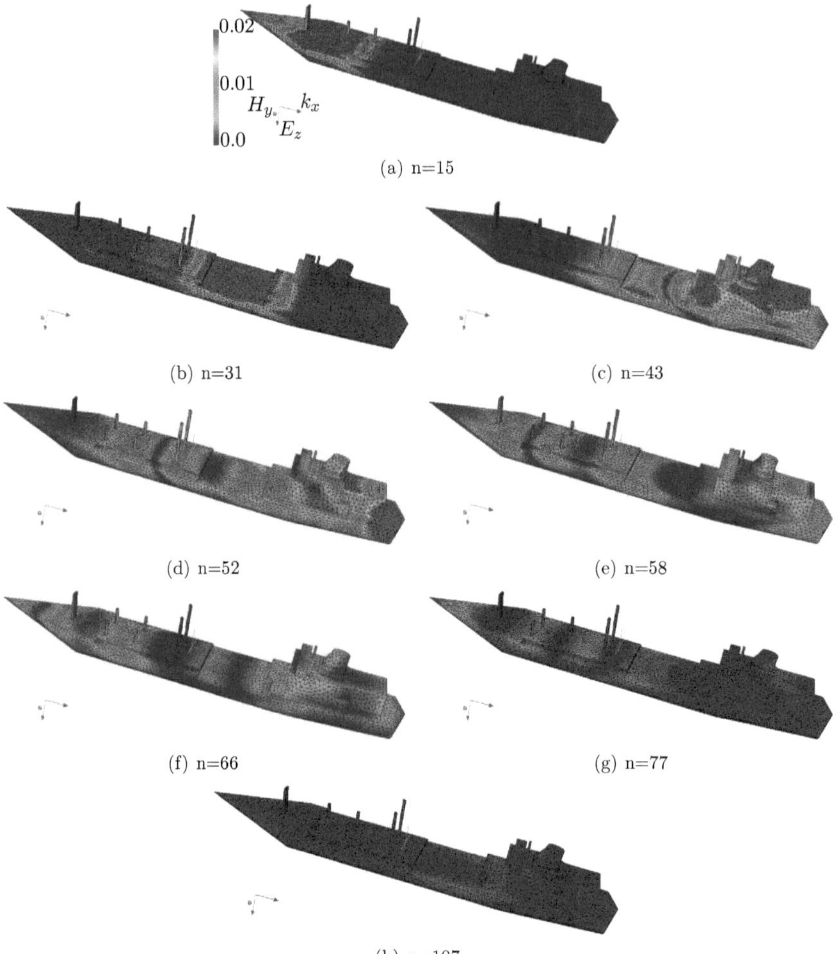

Figure 6.54: The induced surface current magnitude on a 130.8 m-long ship at different time instances $t_n = n\Delta t$ illuminated by a z-polarized Gaussian plane-wave traveling along $\hat{\mathbf{k}} = \hat{\mathbf{x}}$ with the pulse specifications $cT = 40$ m, $ct_0 = 120$ m. The ship is meshed by 10472 triangles and the MFIE is solved for $M = 15708$ degrees of freedoms using the MOT with safety factor $\alpha = 15$ resulting in $c\Delta t = 2.996738$ m. The geometry can be found in CST MICROWAVE STUDIO® 2008 examples. For $n \leq 120$ the solution procedure takes less than half a day and for $M = 32925$ less than two days on a single 64-bit machine with 3 GHz quad-core CPU.

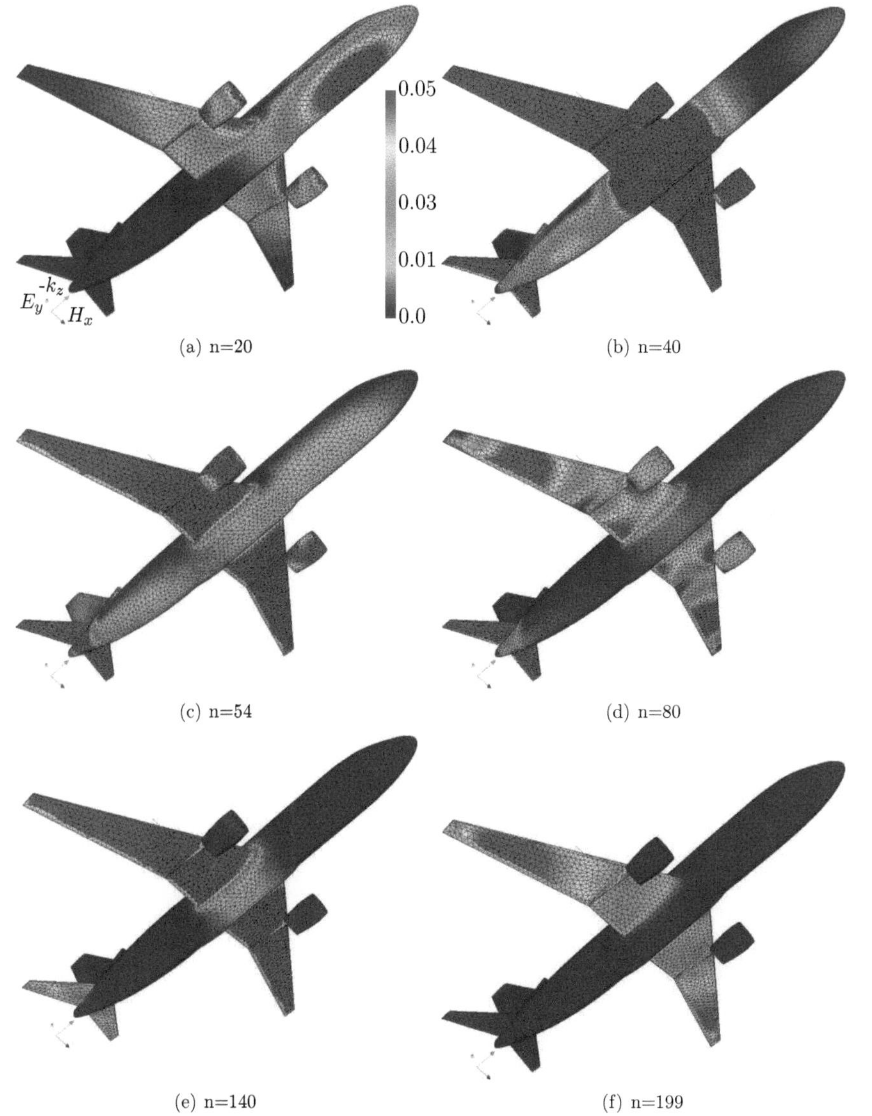

Figure 6.55: The induced surface current magnitude on an Airbus A380 model (123.1915 m-long with 110.7 m distance between the end of wings) at different time instances $t_n = n\Delta t$ illuminated by a y-polarized Gaussian plane-wave traveling along $\hat{\mathbf{k}} = -\hat{\mathbf{z}}$ with the pulse specifications $cT = 80$ m, $ct_0 = 120$ m. The airplane is meshed by 13032 triangles and the CFIE is solved for $M = 19548$ degrees of freedoms using the MOT with safety factor $\alpha = 5 \times 10^5$ resulting in $c\Delta t = 3.620804$ m.

Chapter 7

Summary and Outlook

Diverse time-discretization schemes ensuring the stability of numerical solution of the TDIEs were examined in 3D EM wave scattering problems. First, the influence of time evolutions over space integrals were diminished so as to facilitate the use of extensive research resources existing for spatial discretization in frequency-domain. Detailed formulations for efficient implementation of the stable methods were expressed for alternative forms of the EFIE, MFIE, and CFIE. Adaptive partitioning of source triangles was proposed to automatically guarantee any user-defined desired precision for numerical quadrature routines in calculation of the potential integrals.

The present work approves that the compatibility of the order of the interpolator with that of the integrator has inherent impact on the overall accuracy and stability of numerical results. The use of the implicit (backward) Euler finite difference formula to approximate the time derivatives together with linearly interpolating triangular functions theoretically generate a rigorously stable scheme. In practice, the combination of any of the discussed second order integrators only together with the linear interpolator achieves a non-exploding scheme, that of course in conjunction with the first-order spatial discretization regime totally gives the overall accuracy of the first order for the employed spatio-temporal discretizing method.

As the alternative approach, diverse orders of the time shifted Lagrange polynomials were used to approximate the temporal variation of the current between equally spaced time samples in the BEM. It revealed that smooth Lagrange interpolants enable achievement of stable solution for smaller time steps provided that their closed-form derivatives are utilized as well. The introduced two new quadratic and cubic cardinal B-spline basis functions also improve the extent of the stable region in comparison with the conventional use of the Lagrange polynomials only as the interpolator together with classical difference time integrators.

Entire-domain weighted Laguerre basis functions that are causal, decaying to zero at late-times were used as well to handle the time derivatives and integration analytically. In the MOD approaches, the orthogonality of the bases allows to integrate out the time variable after temporal testing in Galerkin context. The robust, but for long time simulations expensive, FDDM scheme was introduced based-on spectral domain finite-difference approximation. Similarly, the CQM applies a mapping from the Laplace domain to the z-transform domain. The validity of the results were verified through comparison with FIT results.

Although the MOD schemes are yet the only approaches that thoroughly eliminate the late-time instabilities in numerical solution of field integral equations, the high computational expenses preclude their application in large scale scattering problems. A new formulation for the AMOD in conjunction with a parallelizable FFT-based algorithm with complexity of $\mathcal{O}(N_s \log N_s)$ was proposed for accelerating the $\mathcal{O}(N_t)N_t$ retarded matrix-vector multiplies in the numerical solution of the TDIEs. Eventually, owing to the convolutional characteristic of the Toeplitz kernel, the current distribution was efficiently computed by $\mathcal{O}(N_t N_s \log N_s)$ operation cycles per iteration. The method is minimal memory with $\mathcal{O}(N_t N_s)$ storage demands, because only nonredundant entries of the block-Toeplitz matrices are stored. The exterior multi-level Toeplitz arrangement due to the (multipath) periodical extension was incorporated into the deep block-Toeplitz structures relevant to the uniform meshing. This facilitates accurate analysis of large-scale periodic and partially periodic structure with finite size. The techniques presented in this work are directly usable for possible extension to the AIM (precorrected FFT) for fast analysis of irregular cell shapes. The temporal translation invariance can also be exploited for the MOD schemes similar to the spatial shift invariance [139]. Most efficiently, the temporal translation invariance property can be adjoined to the aggregates of spatial Toeplitz matrices in an additional Toeplitz level above all the present levels. Since the algorithm does not benefit from symmetric structures of the matrices, it can also be employed for fast numerical solution of the magnetic (and combined) field integral equations [73]. Nevertheless, research activity on the symmetric multi-level Toeplitz for the EFIE-case is expected.

It was demonstrated that the pointwise sampling on local-support time basis functions is not the only approach yielding the Toeplitz pattern in the marching-on-in-time sequences, rather applying orthogonal entire-domain basis functions with Galerkin testing in time also renders Toeplitz structure along the past order indices. More than six new AMOD schemes were introduced to efficiently eliminate the innermost loop and facilitate the time FFT-based implementations in the MOD methods. The varying-size and fixed-size blocked-aggregates methodologies were adopted to the MOT and analogously CQM and MOD schemes. The latter constellation is unified by the spatial FFT acceleration more easily and is faster only with moderate block size B. It also lessens the redundancy in truncation of the larger blocks filled by zero matrices after N_g steps in the MOT schemes. The hybrid grouping algorithm in Fig. 4.6 is well-suited for large degrees of freedom. It was also shown that the gain of matrix sparsification in wavelet domain using multiresolution wavelet packet transform is not considerable.

The misalignment of the observation normals at the centroid of the test subdomains damages the accurate modeling of cylindrical beam pipes especially in the presence of internal axial excitation. The proposed replacement of the RWG BF by the RT BF over the bodies of translation and revolution improves the accuracy of testing procedure particularly in the excitation of the SIE. Therefore, the efficiency of rectangular cells and the flexibility of triangular cells were combined to simulate open or closed bodies of connected generatrix to irregular shapes more accurately than unitary surface meshing. Numerical results demonstrate that the proposed mixed doublets can reduce the number of unknowns considerably. The advantageous hybrid meshes were employed in the non-dispersive modeling of the long-time propagation of the wake fields in particle accelerator structures, namely when the travelling bunches of charged particles passes through the accelerating cavity with cylindrical tube arms, and the wake potential was calculated as a measure of how the resulting wake fields adversely perturb the motion of charges.

The asymmetric coefficient matrices obtained by conventional quadrature routines impede the usage of iterative matrix solvers for treating the large scale problems by the EFIE.

The introduced simultaneous adaptive refining of quadrature subdomains reduces the condition number of the coefficient matrices and at least halves the computational burden of matrices fill-in process including the retarded ones on the right hand side of the original equation. Nevertheless, the eigenvalue spectrum of the system iteration matrix reveals that only the advanced MOD methods with infinite expansion orders and the CQM with infinite time-frequency samples provide non-dissipative schemes. (October 2010)

Chapter 8

Appendix

8.1 Hilbert Transforms

The Hilbert transform $\hat{x}(t)$ of a time signal $x(t)$ is defined for all times t by the convolution of $x(t)$ with the function $h(t) = \frac{1}{\pi t}$,

$$\hat{x}(t) = \mathcal{H}[x(t)] = P\ V \int_{-\infty}^{\infty} x(\tau)h(t-\tau)\mathrm{d}\tau = \frac{1}{\pi} P\ V \int_{-\infty}^{\infty} \frac{x(\tau)}{t-\tau}\mathrm{d}\tau$$

provided that the integral exists as a principal value. The Fourier transform of the $\hat{X}(\omega)$ of the Hilbert transformed signal $\hat{x}(t)$ is a multiplier operator on $X(\omega)$, the Fourier transform of $x(t)$, that is

$$\hat{X}(\omega) = -j\ \mathrm{sgn}(\omega)\ X(\omega).$$

The Hilbert transform does not change the magnitude of $X(\omega)$. It only causes the phase change of $\mp\frac{\pi}{2}$, namely the positive shift at negative frequencies and vice versa. For bandlimited $X(\omega)$, it can be shown that $\hat{x}(t)$ has exactly the same energy (spectral density) as the (real-valued) $x(t)$. The analytic signal or pre-envelope of $x(t)$ is defined as the complex signal

$$\tilde{x}(t) = x(t) + j\hat{x}(t).$$

The Fourier transform of $\tilde{x}(t)$ is one-sided

$$\tilde{X}(\omega) = \begin{cases} 2X(\omega) & \omega > 0 \\ 0 & \omega < 0. \end{cases}$$

In other words, when the significant frequency content of a bandpass signal $x(t)$ is centered at angular frequency ω_0 and spread over $[\omega_0 - \omega_b, \omega_0 + \omega_b]$ and its mirror $[-\omega_0 - \omega_b, -\omega_0 + \omega_b]$, the major support of $\tilde{X}(\omega)$ has the same bandwidth $2\omega_b$ but lies only in the interval $[\omega_0 - \omega_b, \omega_0 + \omega_b]$. The inverse Fourier transform of the down-shifted passband $\tilde{X}(\omega + \omega_0)$ is called the complex envelope of $x(t)$,

$$\mathcal{X}(t) = \tilde{x}(t)e^{-j\omega_0 t}.$$

Therefore, the original field quantities in Section (2.3.2) can be retrieved from their complex envelopes through

$$x(t) = \mathrm{Re}[\mathcal{X}(t)e^{j\omega_0 t}].$$

8.2 Duffy Transformations

Duffy's trick is an analytical integration technique for singularity cancellation (not extraction) [140]. Assume that the observation point $\mathbf{r} = \mathbf{r}_c$ coincides with the barycenter of a source triangle patch T_k^q with vertices $\mathbf{r}_1, \mathbf{r}_2, \mathbf{r}_3$, and one intends to calculate the scalar potential integral

$$\int_S \frac{1}{R} \mathrm{d}S' = \int_{T_k^q} \frac{1}{|\mathbf{r} - \mathbf{r}'|} \mathrm{d}\mathbf{r}'.$$

First the triangle is partitioned into three subtriangles $T_k^q = \sum_{i=1}^3 T_{k_i}^q$. A local coordinate system (u, v) originating form $\mathbf{r}_c$ is established on two sides of the triangle sharing the free vertex $\mathbf{r}_1$. The variable $\mathbf{r}' = x\hat{\mathbf{x}} + y\hat{\mathbf{y}} + z\hat{\mathbf{z}}$ on the source subdomain can be projected on (u, v) plane

$$\mathbf{r}' = \mathbf{r}_c + (\mathbf{r}_1 - \mathbf{r}_c)u + (\mathbf{r}_2 - \mathbf{r}_1)v = \mathbf{r}_c + \mathbf{u}u + \mathbf{v}v.$$

where the Jacobian of the transformation is double the area of the triangle

$$J_{uv} = \left| \frac{\partial \mathbf{r}'}{\partial u} \times \frac{\partial \mathbf{r}'}{\partial v} \right| = |\mathbf{u} \times \mathbf{v}| = \frac{2}{3} A_k^q$$

and is canceled by the normalization factor of the RWG basis function. By change of variables $u = r$ and $v = rs$, the hypotenuse of the obtained right triangle maps the yet existing singular point on $(u = 0, v = 0)$ to the side of a rectangle on the vertical axis of the new (r, s) coordinate system. The Jacobian determinant

$$J_{rs} = \begin{vmatrix} \frac{\partial u}{\partial r} & \frac{\partial u}{\partial s} \\ \frac{\partial v}{\partial r} & \frac{\partial v}{\partial s} \end{vmatrix} = r$$

then cancels the singularity, i.e.

$$\frac{1}{2A_k^q} \int_{T_{k_1}^q} \frac{1}{|\mathbf{r}' - \mathbf{r}_c|} \mathrm{d}\mathbf{r}' = \int_0^1 \int_0^u \frac{1}{|\mathbf{u}u + \mathbf{v}v|} \mathrm{d}v \, \mathrm{d}u = \int_0^1 \int_0^1 \frac{r}{r|\mathbf{u} + \mathbf{v}s|} \mathrm{d}r \, \mathrm{d}s$$
$$= \int_0^1 \frac{1}{\sqrt{as^2 + 2bs + c}} \mathrm{d}s = \frac{1}{\sqrt{a}} \ln \left(\frac{\sqrt{a(a + 2b + c)} + a + b}{\sqrt{ac} + b} \right). \tag{8.1}$$

where $a = |\mathbf{u}|^2$, $b = \mathbf{u} \cdot \mathbf{v}$, $c = |\mathbf{v}|^2$. Similarly for $i = 2, 3$, integration on $T_{k_i}^q$ one has to only insert $\mathbf{u} = \mathbf{r}_i - \mathbf{r}_c$ and $\mathbf{v} = \mathbf{r}_{i+1} - \mathbf{r}_i$.

The linearly varying vector potential integral can be decoupled into two terms

$$\int_S \frac{\rho}{R} \mathrm{d}S' = \int_{T_k^q} \frac{\mathbf{r}' - \mathbf{r}_i}{|\mathbf{r} - \mathbf{r}'|} \mathrm{d}\mathbf{r}' = \int_{T_k^q} \frac{\mathbf{r}' - \mathbf{r}}{|\mathbf{r}' - \mathbf{r}|} \mathrm{d}\mathbf{r}' + (\mathbf{r} - \mathbf{r}_i) \int_{T_k^q} \frac{1}{|\mathbf{r} - \mathbf{r}'|} \mathrm{d}\mathbf{r}'.$$

in which the scalar integral has been already calculated. When $\mathbf{r} = \mathbf{r}_c$ lies on T_k^q, one can apply the same transformation to the vector integrand.

$$\frac{1}{2A_k^q} \int_{T_{k_i}^q} \frac{\mathbf{r}' - \mathbf{r}_c}{|\mathbf{r}' - \mathbf{r}_c|} \mathrm{d}\mathbf{r}' = \int_0^1 \int_0^u \frac{\mathbf{u}u + \mathbf{v}v}{|\mathbf{u}u + \mathbf{v}v|} \mathrm{d}v \, \mathrm{d}u = \int_0^1 \int_0^1 \frac{r(\mathbf{u} + \mathbf{v}s)}{r|\mathbf{u} + \mathbf{v}s|} r \, \mathrm{d}r \, \mathrm{d}s$$
$$= \frac{\mathbf{u}}{2} \int_0^1 \frac{1}{|\mathbf{u} + \mathbf{v}s|} \mathrm{d}s + \frac{\mathbf{v}}{2} \int_0^1 \frac{s}{|\mathbf{u} + \mathbf{v}s|} \mathrm{d}s$$
$$= \frac{1}{2\sqrt{a}} \left(\mathbf{u} - \mathbf{v}\frac{b}{a} \right) \ln \left(\frac{\sqrt{a(a + 2b + c)} + a + b}{\sqrt{ac} + b} \right) + \frac{1}{2a} \mathbf{v}(\sqrt{a + 2b + c} - \sqrt{c})$$

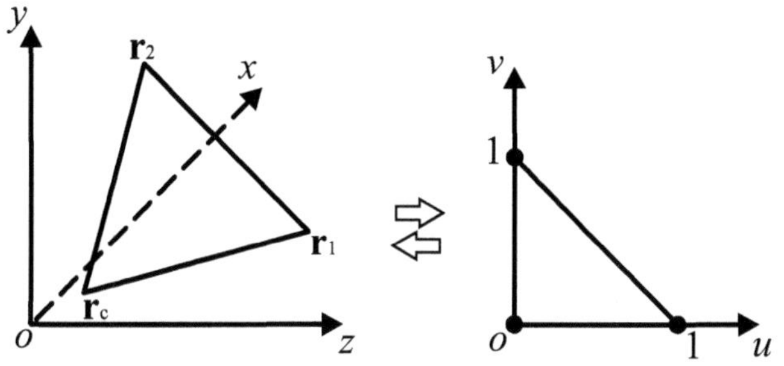

Figure 8.1: Every three slices of the arbitrary source triangle are mapped individually to the unit right triangle in (u,v) plane.

since

$$\int_0^1 \frac{s}{\sqrt{as^2+2bs+c}}\mathrm{d}s = \frac{1}{a}\sqrt{as^2+2bs+c} - \frac{b}{a}\int_0^1 \frac{1}{\sqrt{as^2+2bs+c}}\mathrm{d}s$$

and the last remaining integral has been already evaluated analytically by (8.1). The procedure is repeated similarly for equilateral subdomains $P_k^q = \sum_{i=1}^4 P_{k_i}^q$.

8.3 Inner Products of Vector Bases

The value of (2.50) is zero if the field point does not lie on the source triangle, i.e. $\mathbf{r} \notin T_k^q$, and for self-triangle interactions $(A_m^p = A_k^q)$, it can be calculated analytically as

$$\begin{aligned} F_{mm}^{pp} &= \frac{1}{{A_m^p}^2}\int_{T_m^p} |\rho_m^p|^2 \mathrm{d}S \\ &= \frac{1}{{A_m^p}^2}\int_0^1\int_0^{1-v} |\mathbf{u}u+\mathbf{v}v|^2(2A_m^p)du dv = \frac{|\mathbf{u}|^2+|\mathbf{v}|^2+\mathbf{u}\cdot\mathbf{v}}{6A_m^p} \end{aligned} \quad (8.2)$$

for overlapping edges where $\mathbf{u} = \mathbf{r}_1 - \mathbf{r}_0$, $\mathbf{v} = \mathbf{r}_2 - \mathbf{r}_0$ and $\mathbf{r}_0$ is the position vector of the free vertex of the triangle T_m^p, and

$$\begin{aligned} F_{mk}^{pq} &= \frac{1}{A_m^p A_k^q}\int_{T_m^p} \rho_m^p \cdot \rho'^q_k \mathrm{d}S \\ &= \frac{1}{{A_m^p}^2}\int_0^1\int_0^{1-v} (\mathbf{u}u+\mathbf{v}v)\cdot(\mathbf{u}u+\mathbf{v}v-\mathbf{v})(2A_m^p)du dv \\ &= pq\frac{|\mathbf{u}|^2-|\mathbf{v}|^2-\mathbf{u}\cdot\mathbf{v}}{6A_m^p} \end{aligned} \quad (8.3)$$

for adjacent edges where $\mathbf{r}_1$ is the position vector of the common end of two neighboring edges.

8.4 Laguerre Transform

Consider the set of Laguerre functions of orders j,

$$L_j(t) = \frac{e^t}{j!}\frac{\mathrm{d}^j}{\mathrm{d}t^j}(t^j e^{-t}), \qquad 0 \leq t < \infty, \qquad j = 0, 1, 2, \ldots. \tag{8.4}$$

Since they are orthogonal (3.47), analogous to (3.46), one can expand a causal response function $f(t)$ by the weighted Laguerre polynomials $\phi_j(t)$

$$f(t) = \sum_{j=0}^{\infty} f_j \phi_j(t) \tag{8.5}$$

and define a Laguerre transform

$$\int_0^{\infty} \phi_i(t) f(t) \mathrm{d}t = f_i. \tag{8.6}$$

Fig. 8.2 exhibits the Laguerre transform of four different signals.

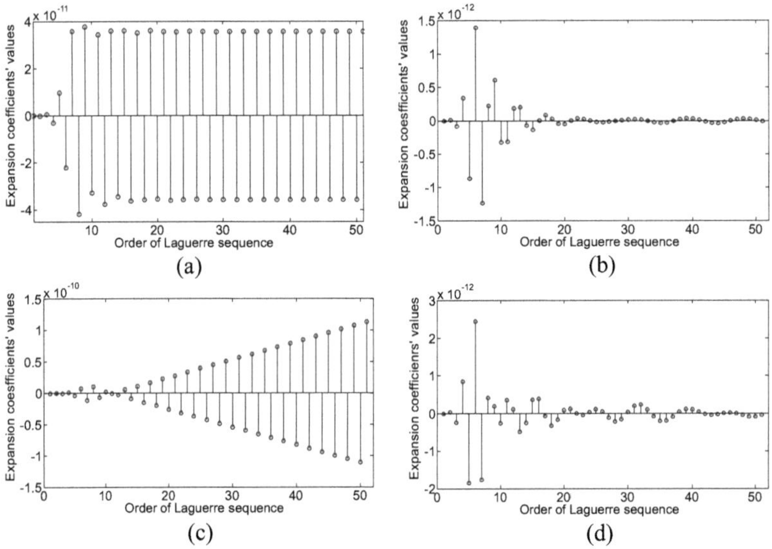

Figure 8.2: Discrete Laguerre transform coefficients f_i for four commonly used excitation pulses. (a) the Gaussian pulse specified in (6.1) and (b) the modulated Gaussian pulsed defined in (6.2). (c) a triangular pulse and (d) a rectangular pulse.

The analytic representation for the derivative of the function $f(t)$ (3.48) is obtained by using the product rule (Leibniz's law) for

$$\int_0^{\infty} \phi_i(t) \frac{\mathrm{d}}{\mathrm{d}t} f(t) \mathrm{d}t = \frac{1}{2} f_i + \sum_{k=0}^{i-1} f_k(t), \tag{8.7}$$

assuming $f(0) = 0$, $\frac{\mathrm{d}}{\mathrm{d}t} f(0) = 0$, and $\phi_i(\infty) = 0$.

8.4.1 Time Weighted Expansion

The definition (3.64) is initiated from

$$I_\nu(\tau) = \int_0^\infty \phi_i(t)\phi_j(t-\tau)\mathrm{d}t \tag{8.8}$$

and simplified through a change of variable $\underline{t} = t - \tau$

$$I_\nu(\tau) = e^{-\frac{\tau}{2}} \int_{-\tau}^\infty e^{-\underline{t}} L_i(\underline{t}+\tau)L_j(\underline{t})\mathrm{d}\underline{t}.$$

The Sheffer's identity reads

$$L_i(\underline{t}+\tau) = \sum_{k=0}^{i} L_k(\underline{t})[L_{i-k}(\tau) - L_{i-k-1}(\tau)]. \tag{8.9}$$

$$I_\nu(\tau) = e^{-\frac{\tau}{2}} \sum_{k=0}^{i} [L_{i-k}(\tau) - L_{i-k-1}(\tau)] \int_{-\tau}^\infty e^{-\underline{t}} L_k(\underline{t})L_j(\underline{t})\mathrm{d}\underline{t}\,.$$

Since the Laguerre functions are defined for $\underline{t} \geq 0$, the lower limit of the integral can be changed from $-\tau$ to 0, and hence, the orthogonality (3.47) renders

$$I_\nu(\tau) = \begin{cases} e^{-\frac{\tau}{2}}[L_{i-j}(\tau) - L_{i-j-1}(\tau)] & j \leq i \\ 0 & j > i. \end{cases} \tag{8.10}$$

8.5 Hermite Transform

Consider the set of Hermite functions of orders j,

$$H_j(t) = (-1)^j e^{t^2} \left(\frac{\partial}{\partial t}\right)^j e^{-t^2}, \qquad -\infty < t < \infty, \qquad j = 0, 1, 2, \ldots. \tag{8.11}$$

Since they are orthogonal (3.104), analogous to (3.46), one can expand the response function $f(t)$

$$f(t) = \sum_{j=0}^\infty f_j h_j(t) \tag{8.12}$$

and define a Hermite transform

$$\int_{-\infty}^\infty h_i(t) f(t)\mathrm{d}t = f_i. \tag{8.13}$$

The analytic representation for the derivative of the function $f(t)$ (3.105) is obtained by using the product rule (Leibniz's law) for

$$\int_{-\infty}^\infty \phi_i(t)\frac{\mathrm{d}}{\mathrm{d}t}f(t)\mathrm{d}t = \sqrt{\frac{j}{2}}f_{j-1} + \sqrt{\frac{j+1}{2}}f_{j+1}, \tag{8.14}$$

assuming $f(0) = 0$ and $h_i(\infty) = 0$.

The definition (3.104) is initiated from

$$I_\nu(\tau) = \int_{-\infty}^{\infty} h_i(t) h_j(t-\tau)\mathrm{d}t \tag{8.15}$$

and simplified through a change of variable $\underline{t} = t - \tau$

$$I_\nu(\tau) = \frac{1}{\sqrt{i!2^i\sqrt{\pi}}}\frac{1}{\sqrt{j!2^j\sqrt{\pi}}} e^{-\frac{\tau^2}{2}} \int_{-\infty}^{\infty} e^{-\underline{t}^2}\, H_i(\underline{t}+\tau) H_j(\underline{t})\mathrm{d}\underline{t}.$$

The Appell's identity reads

$$H_i(\underline{t}+\tau) = \sum_{k=0}^{i} \binom{i}{k} (2\tau)^{i-k} H_k(\underline{t}). \tag{8.16}$$

$$I_\nu(\tau) = \frac{1}{\sqrt{i!2^i\sqrt{\pi}}}\frac{1}{\sqrt{j!2^j\sqrt{\pi}}} e^{-\frac{\tau^2}{2}} \sum_{k=0}^{i} \binom{i}{k} (2\tau)^{i-k} \int_{-\infty}^{\infty} e^{-\underline{t}^2}\, H_k(\underline{t}) H_j(\underline{t})\mathrm{d}\underline{t}\,.$$

and due to the orthogonality (3.104)

$$I_\nu(\tau) = \sqrt{\frac{j!2^j\sqrt{\pi}}{i!2^i\sqrt{\pi}}} e^{-\frac{\tau^2}{2}} \sum_{k=0}^{i} \binom{i}{k} (2\tau)^{i-k} \delta_{kj}.$$

Thus

$$I_\nu(\tau) = \begin{cases} \sqrt{\frac{i!}{j!}}\frac{\sqrt{2^{i-j}}}{(i-j)!} e^{-\frac{\tau^2}{2}} (\tau)^{i-j} & j \le i \\ 0 & j > i. \end{cases} \tag{8.17}$$

8.5.1 Choice of Expansion Order

Assuming that the characterized signal $f(t)$ is bandlimited to frequency B Hertz and it is to be regenerated upto the time duration T, the associated Fourier series are

$$f(t) = \sum_n C_n e^{jn\omega_0 t}$$

where $\omega_0 = \frac{2\pi}{T}$ and since $f(t)$ is a real time signal $C_n^* = C_{-n}$ ($*$ denotes the conjugate transpose). The band-limitedness $-B \le \frac{n}{T} \le B$ implies that the expansion of $f(t)$ has $2BT + 1$ terms,

$$f(t) = \sum_{n=-BT}^{BT} C_n e^{jn\omega_0 t}. \tag{8.18}$$

Therefore, the minimum number of temporal basis function is

$$N = 2BT + 1. \tag{8.19}$$

In non-resonant EM scattering problems, the bandwidth of output depends on the bandwidth of the incident fields. The Fourier transform of the incident Gaussian pulse (6.1) is

$$\bar{\mathbf{E}}^i(\mathbf{r}, f) = \frac{\mathbf{E}_0}{c} e^{-(\frac{\pi T f}{4c})^2} e^{-j2\pi f t_0}. \tag{8.20}$$

Stretching the time interval and the bandwidth of the incident pulse for satisfying a priori set computation precision ϵ,

$$\frac{|\mathbf{E}^i(\mathbf{r}, T)|}{\max |\mathbf{E}^i(\mathbf{r}, t)|} \leq \epsilon \qquad \frac{|\bar{\mathbf{E}}^i(\mathbf{r}, B)|}{\max |\bar{\mathbf{E}}^i(\mathbf{r}, f)|} \leq \epsilon.$$

By neglecting the spatial variance,

$$T = t_0 - \frac{T}{4c} \ln \epsilon \qquad B = -\frac{4c}{\pi T} \ln \epsilon.$$

Thus, the lower bound for the time-bandwidth product

$$N = \frac{8c}{\pi T} \ln \epsilon \left(\frac{T}{4c} \ln \epsilon - t_0 \right) + 1, \tag{8.21}$$

e.g., when $\epsilon = 0.001$ and $cT = 30$ lm in (6.1), $N = 50$ Laguerre or Hermite polynomial orders seem sufficient to approximate the temporal variation of the system response.

8.6 z-Transform

Given a sequence x_n, $n = 0, 1, 2, \ldots$ its z-transform $X(z) = \mathcal{Z}\{x_n\}$ is defined by

$$X(z) = \sum_{n=0}^{\infty} x_n z^{-n}. \tag{8.22}$$

Assuming that the signal so represented is absolutely summable,

$$\sum_{n=0}^{\infty} |x_n| < \infty$$

the inverse z-transform is obtained by

$$x_n = \mathcal{Z}^{-1}\{X(z)\} = \frac{1}{2\pi} \int_{-\pi}^{\pi} X(e^{j\omega}) e^{jn\omega} \mathrm{d}\omega. \tag{8.23}$$

The unit delay property of the z-transform states that for $y_n = x_{n-1}$

$$\mathcal{Z}\{y_n\} = z^{-1} X(z). \tag{8.24}$$

Hence, the difference formula $\mathcal{Z}\{x_n - x_{n-1}\} = (1 - z^{-1}) X(z)$ etc.

8.7 Same Side Technique

function PointinTriangle[$\hat{\mathbf{e}}, \mathbf{r}_1, \mathbf{r}_2, \mathbf{r}_3$]
// The definition of the parameters has been given in Section 5.3
  if
    SameSide[$(e_u, e_v), (u_1, v_1), (u_2, v_2), (u_3, v_3)$] and
    SameSide[$(e_u, e_v), (u_2, v_2), (u_3, v_3), (u_1, v_1)$] and
    SameSide[$(e_u, e_v), (u_3, v_3), (u_1, v_1), (u_2, v_2)$]
  then return true
  else return false
function SameSide[$\mathbf{e}, \mathbf{w}_1, \mathbf{w}_2, \mathbf{w}_3$]
  $\mathbf{p}_1$ = CrossProduct($\mathbf{w}_2 - \mathbf{w}_1,\ \mathbf{e} - \mathbf{w}_1$)
  $\mathbf{p}_2$ = CrossProduct($\mathbf{w}_2 - \mathbf{w}_1, \mathbf{w}_3 - \mathbf{w}_1$)
  if
    DotProduct($\mathbf{p}_1, \mathbf{p}_2$) ≥ 0
  then return true
  else return false

Nomenclature

General Symbols

$*$	Time convolution
$\otimes$	Space convolution
μ	Permeability
ϵ	Permittivity
σ	Conductivity
c	Speed of light
η	Intrinsic impedance
ω	Angular frequency
$\mathbb{R}$	Set of real numbers
S	Surface of the object
$\mathrm{d}S$	Infinitesimal surface area element
D	Maximum linear dimension of the scatterer
$\mathbf{D}_p$	Periodicity displacement
$\hat{\mathbf{D}}$	Periodicity direction
$\hat{n}$	Unit normal surface vector
∇	Gradient
$\nabla\cdot$	Divergence
$\nabla\times$	Curl
$\cdot$	Scalar product
$\times$	Cross product
$< \, , \, >$	Inner product
∂	Differential operator
L^2	Euclidean norm (square summable, Hilbert) space
∞	Infinity
(x, y, z)	Cartesian coordinates
(ρ, ϕ, z)	Cylindrical coordinates
$\forall$	Universal quantifier
$\mathcal{O}(N)$	Order of N
$\lfloor \, \rfloor$	Rounds to the smaller integer
min	Minimum
max	Maximum
u(.)	Unit step function
$\mathrm{I}_1(.)$	First-order modified Bessel function of the first kind
$P\,V$	[Cauchy] principle value
sgn	Sign (signum) function
T	Matrix transpose

List of Notations

t	Time
t_n	Time instance
δt	Time step size
τ	Retarded time
$\hat{\mathbf{r}}$	Unit vector along the radiation direction
$\mathbf{r}$	Source point
$\mathbf{r}'$	Observation point
R	Distance between $\mathbf{r}$ and $\mathbf{r}'$
$\mathbf{E}(\mathbf{r}, t)$	Total electric field strength
$\mathbf{H}(\mathbf{r}, t)$	Total magnetic field strength
$\mathbf{E}^i$	Incident electric field strength
$\mathbf{H}^i$	Incident magnetic field strength
$\mathbf{E}^s$	Scattered electric field strength
$\mathbf{H}^s$	Scattered magnetic field strength
$\mathbf{D}$	Electric flux density
$\mathbf{B}$	Magnetic flux density
$\mathbf{J}(\mathbf{r}, t)$	Electric surface current density
$\tilde{\mathbf{J}}(\mathbf{r}, s)$	Laplace transform of the surface current density
$\mathcal{J}(\mathbf{r}, t)$	Complex envelope of $\mathbf{J}(\mathbf{r}, t)$
$\mathbf{M}(\mathbf{r}, t)$	Magnetic surface current density
$\phi(\mathbf{r}, t)$	Electric scalar potential
$\mathbf{A}(\mathbf{r}, t)$	Magnetic vector potential
$\mathbf{c}(\mathbf{r}, t)$	Hertz vector
$\Phi(\mathbf{r}, t)$	Hertz potential
$\sigma(\mathbf{r}, t)$	Surface charge density
$\mathbf{f}(\mathbf{r})$	Vector basis function
$T(t)$	Time basis function
$\dot{T}(t)$	Time derivative of T
ΔT	Duration of time basis function
L_i	Laguerre polynomial of order i
ϕ_i	Weighted Laguerre polynomial of order i
H_i	Hermite polynomial of order i
h_i	Hermite function of order i
$\mathcal{H}[.]$	Hilbert transform
m	Test subdomain index
k	Source subdomain index
p	Test subdomain side
q	Source subdomain side
n	Time step index
r	Retarded time sample
i	Order step index
j, ι	Testing order index
ν	Order difference
s	Scaling factor

M, N_s	Number of spatial basis functions
N, N_t	Number of time (order) steps or time basis functions
T_m	Triangular patch
P_m	Quadrilateral patch
S_m	Wire segment
ρ_m	Position vector in subdomain
A_m	Subdomains area
l_m	Edge length
e_n	Electric excitation vector
h_n	Magnetic excitation vector
$\bar{\mathrm{V}}_n$	Total excitation vector
$\bar{\bar{\mathrm{Z}}}_0$	Moment (coefficient) matrix
$\bar{\bar{\bar{\mathrm{Z}}}}_r$	Retarded interaction (impedance) matrices
$\bar{\mathrm{Z}}_r$	Array of unique elements in $\bar{\bar{\bar{\mathrm{Z}}}}_r$
$\mathbf{Z}$	Submatrices in $\bar{\bar{\bar{\mathrm{Z}}}}_r$
$I(t)$	Time-varying current coefficients
$\bar{\mathrm{I}}_n$	Present unknown current vector
$\bar{\mathrm{I}}_r$	Past known current vectors
$\underline{\bar{\mathrm{I}}}_r$	Flipped and zero-padded extension of $\bar{\mathrm{I}}_r$
I_ν	Tested Laguerre expansion
I	Identity dyad
$c_{k,j}$	Space-order coefficients
α	Courant safety factor
δ	Interpolation distance
$\delta(t)$	Dirac Delta distribution
δ_{ij}	Kronecker delta
N_x	Number of subdividing grids in horizontal direction
N_y	Number of subdividing grids in vertical direction
N_ϕ	Number of subdividing grids in azimuthal direction
N_z	Number of subdividing grids in longitudinal direction
P	Number of parallel edges along x
Q	Number of parallel edges along y
n_x	Periodic repetitions along x
n_y	Periodic repetitions along y
N_{s0}	Number of bases on unit cell
P	Vector potential polar integration
P	Scalar potential polar integration
Q	Charge of moving bunch particle
F	Force
W	Wake potential function
ε	Relative error / Threshold
σ	Full-width half max of Gaussian pulse
k_0	Wave number in free space
ω_0	Center angular frequency

Acronyms

3D	Three-dimensional
lm	Light meter
ABC	Absorbing Boundary Condition
ACE	Accelerated Cartesian expansions
BF	Basis Functions
CFL	Courant-Friedrichs-Levy
TD	Time Domain
SIE	Surface Integral Equation
TDIE	Time Domain Integral Equation
EFIE	Electric Field Integral Equation
DEFIE	Derivative Form of Electric Field Integral Equation
MFIE	Magnetic Field Integral Equation
CFIE	Combined Field Integral Equation
MOT	Marching-On-in-Time
MOD	Marching-On-in-Degree
MOH	Marching-On-in-orders of Hermite function
MoM	Method of Moments
AMOD	Advanced Marching-On-in-Degree
PEC	Perfect Electric Conductor
PWTD	Plane Wave Time Domain
DFT	Discrete Fourier Transform
FFT	Fast Fourier Transform
AIM	Adaptive Integral Method
ODE	Ordinary Differential Equations
PDE	Partial Differential Equations
DDE	Delay Differential Equations
FIT	Finite Integration Technique
FWT	Fast Wavelet (packet) Transform
FD	Finite Difference
FDTD	Finite Difference Time Domain
FDDM	Finite Difference Delay Modeling
FVM	Finite Volume Method
FEM	Finite Element Method
BEM	Boundary Element Method
EM	Electromagnetic
FSS	Frequency Selective Surface
CAD	Computer Aided Design
CPU	Central Processing Unit
CQM	Convolution Quadrature Method
Radar	Radio Detection and Ranging
RAM	Random Access Memory
RK	Runge-Kutta
RT	Rooftop
RWG	Rao-Wilton-Glisson
TESLA	TeV-Energy Superconducting Linear Accelerator
BCG	Bi-Conjugate Gradient
CSR	Compressed Sparse Row
MKL	Math Kernel Library
PARDISO	Parallel Direct Sparse Solver

Bibliography

[1] S. M. Rao. *Time domain electromagnetics.* Academic Press, San Diego, CA, 1999. 1, 2, 23, 28, 30, 81, 97, 99, 104, 125

[2] M. J. Bluck and S. P. Walker. Time-domain BIE analysis of large three-dimensional electromagnetic scattering problems. *IEEE Trans. Antennas Propag.*, 45(5):894–901, May 1997. 2, 10, 23

[3] B. Shanker, A. A. Ergin, M. Lu, and E. Michielssen. Analysis of transient electromagnetic scattering from closed surfaces using a combined field integral equation. *IEEE Trans. Antennas Propag.*, 48(7):1064–1074, July 2000. 2, 23, 99

[4] B. H. Jung and T. K. Sarkar. Time-domain CFIE for the analysis of transient scattering from arbitrarily shaped 3D conducting objects. *Microw. Opt. Technol. Lett.*, 34(4):289–296, 2002. 2

[5] G. Manara, A. Monorchio, and S. Rosace. A stable time domain boundary element method for the analysis of electromagnetic scattering and radiation problems. *Eng. Anal. Bound. Elements*, 27(4):389–401, April 2003. 2, 23, 30, 81, 82, 104, 110

[6] D. S. Weile, G. Pisharody, N. Chen, B. Shanker, and E. Michielssen. A novel scheme for the solution of the time-domain integral equations of electromagnetics. *IEEE Trans. Antennas Propag.*, 52(1):283–295, January 2004. 2, 19, 23, 32, 33

[7] R. Khlifi and P. Russer. Hybrid space-discretizing method–method of moments for the analysis of transient interface. *IEEE Trans. Microw. Theory Tech.*, 54(12):4440–4447, December 2006. 2, 28, 104, 125

[8] K. Aygün, B. Shanker, A. A. Ergin, and E. Michielssen. A two-level plane wave time-domain algorithm for fast analysis of EMC/EMI problems. *IEEE Trans. Electromagn. Compat.*, 44(1):152–164, February 2002. 2, 30

[9] H. Bagci, A. E. Yilmaz, V. Lomakin, and E. Michielssen. Fast solution of mixed-potential time-domain integral equations for half space environments. *IEEE Trans. Geosci. Remote Sens.*, 43(2):269–279, February 2005. 30

[10] Y.-S. Chung, T. K. Sarkar, B. H. Jung, M. Salazar-Palma, Z. Ji, S. Jang, and K. Kim. Solution of time domain electric field integral equation using Laguerre polynomials. *IEEE Trans. Antennas Propag.*, 52(9):2319–2328, September 2004. 3, 9, 23, 61, 99, 119

[11] J.-L. Hu, C. H. Chan, and Y. Xu. A fast solution of the time-domain integral equation using fast Fourier transformation. *Microw. Opt. Technol. Lett.*, 25(3):172–175, May 2000. 1, 2, 62, 66

[12] S. M. Rao and T. K. Sarkar. An efficient method to evaluate the time-domain scattering from arbitrarily shaped conducting bodies. *Microw. Opt. Technol. Lett.*, 17(5):321–325, April 1998. 2, 60

[13] B. P. Ryne and P. D. Smith. Stability of time marching algorithms for the electric field integral equation. *J. Electromagn. Waves Appl.*, 4:1181–1205, 1990. 2, 125

[14] P. D. Smith. Instabilities in time marching methods for scattering: Cause and rectification. *Electromagn.*, 10:439–451, 1990. 2, 125

[15] S. Dodson, S. P. Walker, and M. J. Bluck. Implicitness and stability of time domain integral equation scattering analysis. *Appl. Comput. Electromagn. Soc. J.*, 13(3):291–301, November 1998.

[16] P. J. Davies and D. B. Duncan. Averaging technique for time-marching schemes for retarded potential integral equations. *Appl. Numer. Math.*, 23(3):291–310, May 1997. 2, 125

[17] T. Abboud, J. C. Nedelec, and J. Volakis. Stable solution of the retarded potential equations. In *Proc. Appl. Comput. Electromagn. Soc. Conf.*, pages 146–151, Monterey, CA, March 2001.

[18] J.-L. Hu and C. H. Chan. Improved temporal basis function for time-domain electric field integral equation method. *Electron. Lett.*, 35(11):883–885, May 1999. 2, 28, 32, 125

[19] J.-L. Hu, C. H. Chan, and Y. Xu. A new temporal basis function for the time-domain integral equation method. *IEEE Microw. Wireless Compon. Lett.*, 11:465–466, November 2001. 2, 3, 28, 32, 125

[20] B. H. Jung and T. K. Sarkar. Time-domain electric-field integral equation with central finite difference. *Microw. Opt. Technol. Lett.*, 31(6):429–435, December 2001.

[21] B. H. Jung, T. K. Sarkar, Y.-S. Chung, M. Salazar-Palma, and Z. Ji. Time-domain combined field integral equation using Laguerre polynomials as temporal basis functions. *Int. J. Numer. Model.: Electron. Netw., Dev. Fields*, 17:251–268, 2004. 3, 9, 23, 107

[22] R. A. Wildman and D. S. Weile. An accurate broad-band method of moments using higher order basis functions and tree-loop decomposition. *IEEE Trans. Antennas Propag.*, 52(11):3005–3011, November 2004. 25

[23] G. X. Jiang, H. B. Zhu, G. Q. Ji, and W. Cao. Improved stable scheme for the time domain integral equation method. *IEEE Microw. Wireless Compon. Lett.*, 17(1):1–3, January 2007. 2, 28, 29, 33, 125

[24] P. Wang, M. Y. Xia, J. M. Jin, and L. Z. Zhou. Time-domain integral equation solvers using quadratic B-spline temporal basis functions. *Microw. Opt. Technol. Lett.*, 49(5):1154–1159, May 2007. 2, 33, 99

[25] J. Pingenot, S. Chakraborty, and V. Jandhyala. Polar integration for exact space-time quadrature in time-domain integral equations. *IEEE Trans. Antennas Propag.*, 54(10):3037–3042, October 2006. 2, 89

[26] B. Shanker, A. A. Ergin, K. Aygün, and E. Michielssen. Analysis of transient electromagnetic scattering phenomena using a two-level plane wave time-domain algorithm. *IEEE Trans. Antennas Propag.*, 48(4):510–523, April 2000. 2

[27] B. Shanker, A. A. Ergin, M. Lu, and E. Michielssen. Fast analysis of transient electromagnetic scattering phenomena using the multilevel plane wave time domain algorithm. *IEEE Trans. Antennas Propag.*, 51(3):628–641, March 2003. 2, 62, 97

[28] M. Vikram and B. Shanker. Fast evaluation of time domain fields in sub-wavelength source/observer distributions using accelerated Cartesian expansions (ACE). *J. Comput. Phys.*, 227(2):1007–1023, December 2007. 2

[29] A. E. Yilmaz, D. S. Weile, J.-M. Jin, and E. Michielssen. A hierarchical FFT algorithm (HIL-FFT) for the fast analysis of transient electromagnetic scattering phenomena. *IEEE Trans. Antennas Propag.*, 50(7):971–982, July 2002. 2, 63

[30] K. Cools, F. P. Andriulli, F. Olyslager, and E. Michielssen. Time domain Calderón identities and their application to the integral equation analysis of scattering by PEC objects Part I: Preconditioning. *IEEE Trans. Antennas Propag.*, 57(8):2352–2364, August 2009. 2, 25

[31] K. Cools, F. P. Andriulli, F. Olyslager, and E. Michielssen. Nullspaces of MFIE and Calderón preconditioned EFIE operators applied to toroidal surfaces. *IEEE Trans. Antennas Propag.*, 57(10):3205–3215, October 2009.

[32] H. Bagci, F. P. Andriulli, K. Cools, F. Olyslager, and E. Michielssen. Calderón multiplicative preconditioner for coupled surface-volume electric field integral equations. *IEEE Trans. Antennas Propag.*, 58(8):2680–2690, August 2010. 2

[33] H. Bagci, F. P. Andriulli, F. Vipiana, G. Vecchi, and E. Michielssen. A well-conditioned integral-equation formulation for efficient transient analysis of electrically small microelectronic devices. *IEEE Trans. Adv. Packag.*, 33(2):468–480, May 2010. 2

[34] A. Bellen and M. Zennaro. *Numerical methods for delay differential equations.* Clarendon Press, Oxford, 2003. 3, 29

[35] A. Geranmayeh, W. Ackermann, and T. Weiland. Proper combination of integrators and interpolators for stable marching-on-in time schemes. In *Proc. 10th IEEE Int. Conf. on Electromag. in Adv. Appl. (ICEAA'07)*, volume 415, pages 948–951, Torino, Italy, September 2007. 3

[36] A. Geranmayeh, W. Ackermann, and T. Weiland. Survey of temporal basis functions for integral equation methods. In *Proc. 7th IEEE Workshop on Computational Electromagnetics in Time-Domain (CEM-TD'07)*, volume 533, pages 1–4, Perugia, Italy, October 2007. 3, 34

[37] S. Balasubramanian, S. N. Lalgudi, and B. Shanker. Fast-integral-equation scheme for computing magnetostatic fields in nonlinear media. *IEEE Trans. Mag.*, 38(5):3426–3432, September 2002. 5

[38] G. Kobidze and B. Shanker. Integral equation based analysis of scattering from 3-D inhomogeneous anisotropic bodies. *IEEE Trans. Antennas Propag.*, 52(10):2650–2658, October 2004. 5

[39] M. Clemens and T. Weiland. Discrete electromagnetism with the finite integration technique. *Progress In Electromagnetics Research*, 32:65–87, 2001. 5, 99

[40] P. M. Morse and H. Feshbach. *Methods of Theoretical Physics.* McGraw-Hill, New York, 1953. 6

[41] Z. Ji, T. K. Sarkar, B. H. Jung, Y.-S. Chung, M. Salazar-Palma, and M. Yuan. A stable solution of time domain electric field integral equation for thin-wire antennas using the Laguerre polynomials. *IEEE Trans. Antennas Propag.*, 52(10):2641–2649, October 2004. 9

[42] B. H. Jung, T. K. Sarkar, Y.-S. Chung, M. Salazar-Palma, and Z. Ji. Analysis of transient electromagnetic scattering from dielectric objects using a combined-field integral equation. *Microw. Opt. Technol. Lett.*, 40(6):476–481, March 2004. 9

[43] Z. Ji, T. K. Sarkar, B. H. Jung, M. Yuan, and M. Salazar-Palma. Solving time domain electric field integral equation without the time variable. *IEEE Trans. Antennas Propag.*, 54(1):258–262, January 2006. 9, 41, 119

[44] J. V. Bladel. *Electromagnetic Fields.* Hemisphere Publishing Corporation, Washington DC, 1985. 10

[45] H. Kawaguchi. Time-domain analysis of electromagnetic wave fields by boundary integral equation method. *Eng. Anal. Bound. Elements*, 27(4):291–304, April 2003. 10, 140

[46] D. S. Jones. *Methods in Electromagnetic Wave Propagation.* Oxford Sci, Oxford, UK, 1994. 10

[47] A. Mohan and D. S. Weile. A hybrid method of moments-marching on in time method for the solution of electromagnetic scattering problems. *IEEE Trans. Antennas Propag.*, 53(3):1237–1242, March 2005. 12

[48] D. R. Wilton, S. M. Rao, A. W. Glisson, D. H. Schaubert, O. M. Al-Bundak, and C. M. Butler. Potential integrals for uniform and linear source distributions on polygonal and polyhedral domains. *IEEE Trans. Antennas Propag.*, (3):276–281, March 1984. 13, 82, 83

[49] R. F. Harrington. *Field computation by the moment methods.* IEEE Press, New York, 1993. 14

[50] B. M. Kolundzija and A. R. Djordjevic. *Electromagnetic modeling of composite metallic and dielectric structures: The theorotical background of WIPL-D software.* Artech House, Norwood, MA, 2002. 14, 15, 17

[51] I. Hänninen, M. Taskinen, and J. Sarvas. Singularity subtraction integral formulae for surface integral equations with RWG, rooftop and hybrid basis functions. *Progress In Electromagnetics Research*, 63:243–278, 2006. 14, 15

[52] W. Cai, T. Yu, H. Wang, and Y. Yu. High-order mixed RWG basis functions for electromagnetic applications. *IEEE Trans. Microw. Theory Tech.*, 49(7):1295–1303, July 2001. 14, 15, 17

[53] Z. Zeng and C. C. Lu. Discretization of hybrid VSIE using mixed mesh elements with zeroth-order Galerkin basis functions. *IEEE Trans. Antennas Propag.*, 54(6):1863–1870, June 2006. 15

[54] IE3D v14.0. Technical report, Zeland Software, Inc., Fremont, CA, 2008. 15

[55] H. Kawaguchi. Stable time-domain boundary integral equation method for axisymmetric coupled charge-electromagnetic field problems. *IEEE Trans. Mag.*, 38(2):749–752, March 2002. 17, 109, 140

[56] CST STUDIO SUITE ™ 2009. Technical report, CST GmbH, Darmstadt, Germany, 2009. 18, 99, 145

[57] L. C. Trintinalia and H. Ling. First order triangular patch basis functions for electromagnetic scattering analysis. *J. Electromagn. Waves Appl.*, 15(11):1521–1537, November 2001. 18

[58] L. C. Trintinalia and H. Ling. Integral equation modeling of multilayered doubly-periodic lossy structures using periodic boundary condition and a connection scheme. *IEEE Trans. Antennas Propag.*, 52(9):2253–2261, September 2004. 18

[59] Ö. Ergül and L. Gürel. Linear-linear basis functions for MLFMA solutions of magnetic-field and combined-field integral equations. *IEEE Trans. Antennas Propag.*, 55(4):1103–1110, April 2007. 18

[60] Ö. Ergül and L. Gürel. The use of curl-conforming basis functions for the magnetic-field integral equation. *IEEE Trans. Antennas Propag.*, 54(7):1917–1926, July 2006. 18

[61] A. I. Mackenzie, M. E. Baginski, and S. M. Rao. New basis functions for the electromagnetic solution of arbitrarily-shaped, three dimensional conducting bodies using method of moments. *Microw. Opt. Technol. Lett.*, 50(4):1121–1124, April 2008. 18

[62] A. I. Mackenzie, S. M. Rao, and M. E. Baginski. Electromagnetic scattering from arbitrarily shaped dielectric bodies using paired pulse vector basis functions and method of moments. *IEEE Trans. Antennas Propag.*, 57(7):2076–2083, July 2009. 18

[63] N.-W. Chen. A magnetic frill source model for time-domain integral-equation based solvers. *IEEE Trans. Antennas Propag.*, 55(11):3093–3098, November 2007. 18

[64] M. Abramowitz and I. A. Stegun. *Handbook of Mathematical Functions with Formulas, Graphs, and Mathematical Tables.* Dover, New York, 1972. 20, 28, 57, 58, 91

[65] E. Bleszynski, M. Bleszynski, and T. Jaroszewicz. A new fast time domain integral equation solution algorithm. In *Proc. IEEE Antennas Propag. Soc. Int. Symp.*, volume 4, pages 176–180, Boston, MA, July 2001. 23, 62

[66] G. Pisharody and D. S. Weile. Robust solution of time-domain integral equations using loop-tree decomposition and bandlimited extrapolation. *IEEE Trans. Antennas Propag.*, 53(6):2089–2098, June 2005. 25, 33, 99

[67] R. D. Graglia, D. R. Wilton, and A. F. Peterson. Higher order interpolatory vector bases for computational electromagnetics. *IEEE Trans. Antennas Propag.*, 45(3):329–342, March 1997. 25

[68] D. Jiao, A. A. Ergin, B. Shanker, J.-M. Jin, and E. Michielssen. A fast higher-order time-domain finite element-boundary integral method for 3-D electromagnetic scattering analysis. *IEEE Trans. Antennas Propag.*, 50(9):1192–1202, September 2002. 25

[69] R. Adams and N. J. Champagne. A numerical implementation of a modified form of the electric field integral equation. *IEEE Trans. Antennas Propag.*, 52(9):2262–2266, September 2004. 25

[70] Y. Wang and T. Itoh. Envelope-Finite-Element (EVFE) technique – a more efficient time-domain scheme. *IEEE Trans. Microw. Theory Tech.*, 49(12):2241–2247, December 2001. 27

[71] S. E. Bayer and A. A. Ergin. A stable marching-on-in-time scheme for wire scatterers using a Newmark-Beta formulation. *Progress In Electromagnetics Research B*, 6:337–360, 2008. 28

[72] W. H. Press, S. A. Teukolsky, W. T. Vetterling, and B. P. Flannery. *Numerical Recipes: the Art of Scientific Computing, 3rd ed.* Cambridge Univ. Press, Cambridge, 2007. 34

[73] A. Geranmayeh, W. Ackermann, and T. Weiland. Temporal discretization choices for stable boundary element methods in electromagnetic scattering problems. *Appl. Numer. Math.*, 59(11):2751–2773, November 2009. 43, 61, 160

[74] J. Lacik and Z. Raida. Modeling microwave structures in time domain using Laguerre polynomials. *Radioengineering*, 15(3):1–9, September 2006. 44

[75] B. G. Mikhailenko. Spectral Laguerre method for the approximate solution of time dependent problems. *Appl. Math. Lett.*, 12(4):105–110, May 1999. 45

[76] Y. Shi and R. S. Chen. Analysis of transient electromagnetic scattering using Hermite polynomials. In *Proc. IEEE Antennas Propag. Soc. Int. Symp.*, pages 3895–3898, Albuquerque, NM, July 2006. 47

[77] J. Lacik and Z. Raida. On using orthogonal polynomials for transient analysis of wire antennas. In *Proc. 14th Conf. on Microwave Tech. (COMITE'08)*, pages 1–4, Prague, Czech, April 2008. 47

[78] Ch. Lubich and R. Schneider. Time discretization of parabolic boundary integral equations. *Num. Math.*, 63(1):455–481, December 1992. 49

[79] X. Wang, R. A. Wildman, D. S. Weile, and P. Monk. A finite difference delay modeling approach to the discretization of the time domain integral equations of electromagnetics. *IEEE Trans. Antennas Propag.*, 56(8):2442–2452, August 2008. 49, 51

[80] W. Hackbusch, W. Kress, and S. A. Sauter. Sparse convolution quadrature for time domain boundary integral formulations of the wave equation. *IMA J. Numer. Anal.*, 29(1):158–179, March 2009. 49, 51

[81] L. Kielhorn and M. Schanz. Convolution quadrature method-based symmetric Galerkin boundary element method for 3-d elastodynamics. *Int. J. Numer. Meth. Engng.*, 76(11):1724–1746, June 2008. 52

[82] L. Banjai and S. A. Sauter. Rapid solution of the wave equation in unbounded domains. *SIAM J. Numer. Anal.*, 47(1):227–249, October 2008. 52, 53

[83] A. Geranmayeh, W. Ackermann, and T. Weiland. Finite difference delay modeling of potential time integrals. In *Proc. IEEE Antennas Propag. Soc. Int. Symp.*, number 11, pages 1–4, Toronto, ON, 2010. 54

[84] X. Wang and D. S. Weile. Electromagnetic scattering from dispersive dielectric scatterers using the finite difference delay modeling method. *IEEE Trans. Antennas Propag.*, 58(5):1720–1730, May 2010. 55

[85] X. Wang and D. S. Weile. Finite difference delay modeling with Runge-Kutta methods for the discretization of time domain integral equations. In *Proc. IEEE Antennas Propag. Soc. Int. Symp.*, volume 312, pages 1–4, Toronto, ON, 2010. 55

[86] W. C. Chew. A new look at reciprocity and energy conservation theorems in electromagnetics. *IEEE Trans. Antennas Propag.*, 56(4):970–975, April 2008. 56, 127

[87] W. C. Chew, M. S. Tong, and B. Hu. *Integral equation methods for electromagnetic and elastic waves.* Morgan & Claypool, USA, 2009. 56

[88] D. Sievers, T. F. Eibert, and V. Hansen. Correction to "on the calculation of potential integrals for linear source distributions on triangular domains". *IEEE Trans. Antennas Propag.*, 53(9):3113, September 2005. 58

[89] P. Arcioni, M. Bressan, and L. Perregrini. On the evaluation of the double surface integrals arising in the application of the boundary integral method to 3-D problems. *IEEE Trans. Microw. Theory Tech.*, 45(3):436–439, March 1997. 58

[90] A. E. Yilmaz, J.-M. Jin, and E. Michielssen. A parallel FFT accelerated transient field-circuit simulator. *IEEE Trans. Microw. Theory Tech.*, 53(9):2851–2865, September 2005. 61, 62

[91] W. Yu, D. Fang, and C. Zhou. Marching-on-in-degree based time-domain magnetic field integral equation method for bodies of revolution. *IEEE Microw. Wireless Compon. Lett.*, 17(12):813–815, December 2007. 61

[92] B. H. Jung, Z. Ji, T. K. Sarkar, M. Salazar-Palma, and M. Yuan. A comparison of marching-on in time method with marching-on in degree method for the TDIE solver. *Progress In Electromagnetics Research*, 70:281–296, 2007. 61

[93] A. E. Yilmaz, D. S. Weile, B. Shanker, J.-M. Jin, and E. Michielssen. Fast analysis of transient scattering in lossy media. *IEEE Antennas Wireless Propag. Lett.*, 1:14–17, 2002. 62

[94] A. Geranmayeh, W. Ackermann, and T. Weiland. Space-FFT-accelerated marching-on-in-degree methods for finite periodic structures. *Int. J. Microwave Wireless Technol.*, 1(4):331–337, August 2009. 62

[95] A. E. Yilmaz, D. S. Weile, J.-M. Jin, and E. Michielssen. A fast Fourier transform accelerated marching-on-in-time algorithm for electromagnetic analysis. *Electromagnetics*, 21(3):181–197, April 2001. 62, 65

[96] E. Bleszynski, M. Bleszynski, and T. Jaroszewicz. Block-Toeplitz fast integral equation solver for large finite periodic and partialy periodic antenna arrays. In *IEEE Topical Conference on Wireless Communication Technology*, pages 428–429, Honolulu, HI, October 2003. 62

[97] B. E. Barrowes, F. L. Teixeira, and J. A. Kong. Fast algorithm for matrix-vector multiply of asymmetric multilevel block-Toeplitz matrices in 3-D scattering. *Microw. Opt. Technol. Lett.*, 31(1):28–32, October 2001. 62, 66, 68, 76

[98] A. E. Yilmaz, J.-M. Jin, and E. Michielssen. Time domain adaptive integral method for surface integral equations. *IEEE Trans. Antennas Propag.*, 52(10):2692–2708, October 2004. 62, 68

[99] T. H. Ng, T. T. Chia, and B. L. Ooi. Alternative efficient scheme for the computation of aggregate block Toeplitz matrix vector multiply in TD-AIM. In *24th Ann. Rev. Prog. Appl. Comp. Electromagn. (ACES'08)*, pages 640–645, Niagara Falls, Canada, April 2008. 63, 73

[100] W. B. Lu, T. J. Cui, Z. G. Qian, X. X. Yin, and W. Hong. Accurate analysis of large-scale periodic structures using an efficient sub-entire-domain basis function method. *IEEE Trans. Antennas Propag.*, 52(11):3078–3085, November 2004. 67

[101] D. Sievers. *Anwendung finiter Gruppen zur effizienten Berechnung elektromagnetischer Felder in symmetrischen Strukturen auf Basis der Randelementmethode.* PhD thesis, Technische Universität Darmstadt, Institut für Theorie Elektromagnetischer Felder (TEMF), Schlossgartenstrasse 8, 64289 Darmstadt, Germany, 2008. 68, 82

[102] H. Bagci, A. E. Yilmaz, and E. Michielssen. An FFT-accelerated time-domain multiconductor transmission line simulator. *IEEE Trans. Electromagn. Compat.*, 52(1):199–214, February 2010. 74

[103] F. P. Andriulli, H. Bagci, F. Vipiana, G. Vecchi, and E. Michielssen. A marching-on-in-time hierarchical scheme for the solution of the time domain electric field integral equation. *IEEE Trans. Antennas Propag.*, 55(12):3734–3738, December 2007. 76

[104] G. W. Pan. *Wavelets in electromagnetics and device modeling.* John Wiley, New Jersey, 2003. 76

[105] C. K Chui. *Wavelets: a tutorial in theory and applications.* Academic Press, New York, 1992. 76, 77

[106] W. Dahmen. Multiscale and wavelet methods for operator equations. *Multiscale Problems and Methods in Numerical Simulations*, 18(25):31–96, August 2003. 76

[107] I. Daubechies. *Ten lectures on wavelets.* SIAM, Philadelphia, PA, 1992. 77

[108] E. Bleszynski, M. Bleszynski, and T. Jaroszewicz. AIM: Adaptive integral method for solving large-scale electromagnetic scattering and radiation problems. *Radio Sci.*, 31(5):1225–1251, September 1996. 78

[109] H. T. Anastassiu, M. Smelyanskiy, S. S. Bindiganavale, and J. L. Volakis. Scattering from relatively flat surfaces using the adaptive integral method. *Radio Sci.*, 33(1):7–16, January 1998. 79

[110] S. S. Bindiganavale, J. L. Volakis, and H. T. Anastassiu. Scattering from planar structures containing small features using the adaptive integral method (AIM). *IEEE Trans. Antennas Propag.*, 46(12):1867–1878, December 1998. 79

[111] V. I. Okhmatovski, J. D. Morsey, and A. C. Cangellaris. Loop-tree implementation of the adaptive integral method (AIM) for numerically-stable, broadband, fast electromagnetic modeling. *IEEE Trans. Antennas Propag.*, 52(8):2130–2140, 2004. 79

[112] S. M. Seo, C.-F. Wang, and J.-F. Lee. Analyzing PEC scattering structure using an IE-FFT algorithm. *Appl. Comput. Electromagn. Soc. J.*, 24(2):116–128, April 2009. 79

[113] O. Bakir, H. Bagci, and E. Michielssen. Adaptive integral method with fast Gaussian gridding for solving combined field integral equations. *Waves Rand. Complex Media*, 19(1):147–161, February 2009. 79

[114] T. F. Eibert and V. Hansen. On the calculation of potential integrals for linear source distributions on triangular domains. *IEEE Trans. Antennas Propag.*, 43(12):1499–1502, December 1995. 82

[115] A. Tzoulis and T. F. Eibert. Review of singular potential integrals for method of moments solutions of surface integral equations. *Adv. Radio Sci.*, 2:93–99, May 2004. 82

[116] P. Ylä-Oijala and M. Taskinen. Calculation of CFIE impedance matrix elements with RWG and $n\times$RWG functions. *IEEE Trans. Antennas Propag.*, 51(8):1837–1846, August 2003. 82, 83, 87

[117] R. D. Graglia. On the numerical integration of the linear shape functions times the 3-D Green's function or its gradient on a plane triangle. *IEEE Trans. Antennas Propag.*, 41(10):1448–1455, October 1993. 83

[118] T. Ha-Duong. *On retarded potential boundary integral equations and their discretisation.* Topics in Computational Wave Propagation: Direct and Inverse Problems, Berlin: Springer-Verlag, m. ainsworth, p. davies, d. duncan, p. martin, and b. rynne, eds. edition, 2003. 84

[119] B. Shanker, M. Lu, J. Yuan, and E. Michielssen. Time domain integral equation analysis of scattering from composite bodies via exact evaluation of radiation fields. *IEEE Trans. Antennas Propag.*, 57(5):1506–1520, May 2009. 84, 86

[120] M. Lu and E. Michielssen. Closed form evaluation of time domain fields due to Rao-Wilton-Glisson sources for use in marching-on-in-time based EFIE solvers. In *Proc. IEEE Antennas Propag. Soc. Int. Symp.*, volume 1, pages 74–77, San Antonio, TX, June 2002. 84

[121] Y. shi, M.-Y. Xia, R.-S. Chen, E. Michielssen, and M. Lu. A stable marching-on-in-time solver for time domain surface electric field integral equations based on exact integration technique. In *Proc. IEEE Antennas Propag. Soc. Int. Symp.*, volume 312, pages 1–4, Toronto, ON, 2010. 87, 88

[122] S. D. Gedney, A. Zhu, and C.-C. Lu. Study of mixed-order basis functions for the locally corrected Nyström method. *IEEE Trans. Antennas Propag.*, 52(11):2996–3004, November 2004. 93

[123] R. A. Wildman and D. S. Weile. Two-dimensional transverse-magnetic time-domain scattering using the Nyström method and bandlimited extrapolation. *IEEE Trans. Antennas Propag.*, 53(7):2259–2266, July 2005. 93

[124] H. A. Ülkü and A. A. Ergin. Analytical evaluation of transient magnetic fields due to RWG current bases. *IEEE Trans. Antennas Propag.*, 55(12):3565–3575, December 2007. 95

[125] K. Fujita, H. Kawaguchi, T. Weiland, and S. Tomioka. Three-dimensional wake field computations based on scattered-field time domain boundary element method. *IEEE Trans. Nucl. Sci.*, 56(4):2341–2350, August 2009. 117

[126] O. Podebrad, M. Clemens, and T. Weiland. New flexible subgridding scheme for the finite integration technique. *IEEE Trans. Mag.*, 39(3):1662–1665, May 2003. 122

[127] A. Geranmayeh, W. Ackermann, and T. Weiland. Symplectic time integration methods for retarded potential integral equations. In *Proc. IEEE Antennas Propag. Soc. Int. Symp.*, volume 328, pages 1–4, Charleston, SC, June 2009. 123

[128] I. Zagorodnov, R. Schuhmann, and T. Weiland. Long-time numerical computation of electromagnetic fields in the vicinity of a relativistic source. *J. Comput. Phys.*, 191(2):525–541, November 2003. 140

[129] C. K. Birdsall and A. B. Langdon. *Plasma physics via computer simulation (hardcover).* Taylor & Francis, London, 1991. 140

[130] K. Fujita, H. Kawaguchi, I. Zagorodnov, and T. Weiland. Time domain wake field computation with boundary element method. *IEEE Trans. Nucl. Sci.*, 53(2):431–439, April 2006. 140

[131] K. Fujita, H. Kawaguchi, S. Nishiyama, S. Tomioka, T. Enoto, I. Zagorodnov, and T. Weiland. Scattered-field time domain boundary element method and its application to transient electromagnetic field simulation in particle accelerator physics. *IEICE Trans. Electron.*, (2):265–274, February 2007. 140

[132] K. Fujita, H. Kawaguchi, R. Hampel, W. F. O. Müller, T. Weiland, and S. Tomioka. Time domain boundary element analysis of wake fields in long accelerator structures. *IEEE Trans. Nucl. Sci.*, 55(5):2584–2591, October 2008. 140

[133] O. Napoly, Y. H. Chin, and B. Zotter. A generalized method for calculating wake potentials. In *Proc. Part. Accel. Conf.*, volume 5, pages 3447–3449, Washington, DC, May 1993. 144

[134] T. Weiland. Transverse beam cavity interaction. part i: short range forces. *Nucl. Instr. Meth. Phys. Res. A*, 212(1-3):13–22, July 1983. 145

[135] Y. H. Chin. User's guide for ABCI version 8.8 (azimuthal beam cavity interaction). Technical report, LBL-35258, UC-414, CERN, Switzerland, February 1994. 145

[136] I. N. Onishchenko, D. Yu. Sidorenko, and G. V. Sotnikov. Structure of electromagnetic field excited by an electron bunch in a semi-infinite dielectric-filled waveguide. *Phys. Rev. E*, 65(6):066501–066511, June 2002. 145, 148

[137] A. Geranmayeh, W. Ackermann, and T. Weiland. Absorbing boundary edge elements for time-domain surface integral equations. In *U.R.S.I. Landesausschuss in der Bundesrepublik Deutschland e.V. (Kleinheubacher Tagung'10)*, number 1547, page 1, Miltenberg, Germany, October 2010. 145

[138] O. Napoly, Y. H. Chin, and B. Zotter. A generalized method for calculating wake potentials. *Nucl. Instr. Meth. Phys. Res. A*, 334(2-3):255–265, October 1993. 145

[139] A. Geranmayeh, W. Ackermann, and T. Weiland. Toeplitz property on order indices of Laguerre expansion methods. In *IEEE MTT-S Int. Microwave Symp. (IMS'09) Dig.*, volume TU4G, pages 253–256, Boston, MA, June 2009. 160

[140] M. G. Duffy. Quadrature over a pyramid or cube of integrands with a singularity at a vertex. *SIAM J. Numer. Anal.*, 19(6):1260–1262, December 1982. 164

[141] A. Abubakar and P. M. van den Berg. Iterative forward and inverse algorithms based on domain integral equations for three-dimensional electric and magnetic objects. *J. Comput. Phys.*, 195(1):236–262, March 2004.

[142] Z. Liu, R. J. Adams, and L. Carin. Well-conditioned MLFMA formulation for closed PEC targets in the vicinity of a half space. *IEEE Trans. Antennas Propag.*, 51(10):2822–2829, October 2003.

[143] R. J. Adams, X. Yuan, X. Xin, J.-S. Choi, S. D. Gedney, and F. X. Canning. A numerical implementation of a modified form of the electric field integral equation. *IEEE Trans. Antennas Propag.*, 56(8):2427–2441, August 2008.

[144] A. Bellen and M. Zennaro. Stability properties of interpolants for runge-kutta methods. *SIAM J. Numer. Anal.*, 25:411–432, April 1988.

[145] G. Manara, A. Monorchio, and R. Reggiannini. A space-time discretization criterion for a stable time-marching solution of the electric field integral equation. *IEEE Trans. Antennas Propag.*, 45(3):527–532, March 1997.

[146] B. H. Jung, T. K. Sarkar, and Y.-S. Chung. A survey of various frequency domain integral equations for the analysis of scattering from three-dimensional dielectric objects. *Progress In Electromagnetics Research*, 36:193–246, 2002.

Biography

Amir Geranmayeh was born in 1980, in Tehran, Iran. He received the B.S. degree in electrical engineering and M.S. degree (with distinction) in telecommunication engineering both from Amirkabir University of Technology (Tehran Polytechnic), in 2002 and 2005, respectively. Since 2005, he is a Ph.D. candidate in Electrical Engineering and Information Technology at Darmstadt University of Technology, Darmstadt, Germany.

Between 2001 and 2005, he served as a research assistant at national center of excellence in radio communications and power engineering, Tehran Polytechnic. In 2002 and 2005, he was a research engineer at the Iran Telecommunication Research Center (ITRC) and Niroo Research Institute, the leading research organizations of Iran Ministry of Information & Communication Technology and Ministry of Energy, respectively. In late 2005, he joined the Institut für Theorie Elektromagnetischer Felder, Technische Universität Darmstadt. His primary research interest is applied computational electromagnetics with emphasis on integral equation methods.

Dr. Geranmayeh was the recipient of full Deutschen Forschungsgemeinschaft (DFG) graduate research fellowship, honourably mentioned IEEE AP-S'09 student paper award, the IMS'09 and the European Microwave Association special travel grants, and the ITRC's grants for both the B.S. and M.S. theses. He was also the 2nd nominee for the EuMC'08 young engineer prizes and the finalist of IEEE/ACES'05 student paper contest. He had an invited talk for Graduiertenkolleg "Physik und Technik von Beschleunigern" in summer 2005 and since then he has reviewed for more than ten scientific and technical periodicals, such as three times for *IEEE Trans. Electromagn. Compat.*

MIX
Papier aus verantwortungsvollen Quellen
Paper from responsible sources
FSC® C105338

Printed by Books on Demand GmbH, Norderstedt / Germany